Inventing Global Ecology

Ohio University Press
Series in Ecology and History

James L. A. Webb, Jr., Series Editor

Conrad Totman,
The Green Archipelago: Forestry in Preindustrial Japan

Timo Myllyntaus and Mikko Saiku, eds.,
Encountering the Past in Nature: Essays in Environmental History

James L. A. Webb, Jr.,
Tropical Pioneers: Human Agency and Ecological Change in the
Highlands of Sri Lanka, 1800–1900

Stephen Dovers, Ruth Edgecombe, and Bill Guest, eds.,
South Africa's Environmental History: Cases and Comparisons

David M. Anderson,
Eroding the Commons: The Politics of Ecology in Baringo, Kenya,
1890s–1963

William Beinart and JoAnn McGregor, eds.,
Social History and African Environments

Michael L. Lewis,
Inventing Global Ecology: Tracking the Biodiversity Ideal in India,
1947–1997

Inventing Global Ecology

Tracking the Biodiversity Ideal
in India, 1947–1997

Michael L. Lewis

OHIO UNIVERSITY PRESS

ATHENS

Ohio University Press, Athens, Ohio 45701
© 2004 by Ohio University Press

Printed in the United States of America
All rights reserved

First published in India by Orient Longman Private Limited, and in the UK by
Sangam Books Limited, UK, as *Inventing Global Ecology*.

Ohio University Press books are printed on acid-free paper ∞ ™

12 11 10 09 08 07 06 05 04   5 4 3 2 1

Library of Congress Cataloging-in-Publication Data

Lewis, Michael L., 1971–
  Inventing global ecology : tracking the biodiversity ideal in India,
1947–1997 / Michael L. Lewis.—1st U.S. ed.
    p. cm. — (Ohio University Press series in ecology and history)
  Originally published: Hyderabad : Orient Longman, 2003.
  Includes bibliographical references and index.
  ISBN 0-8214-1540-9 (cloth : alk. paper) — ISBN 0-8214-1541-7 (pbk.
: alk. paper)
    1. Biological diversity conservation—India. 2. Biological diversity—
Monitoring—India. 3. Animal ecology—Research—India. I. Title. II.
Series.
  QH77.I4L48 2004
  333.95'16'0954--dc22

                                                    2004002720

Contents

Illustrations

Maps

Figures

Abbreviations

AID	Agency for International Development
BCI	Barro Colorado Island
BNHS	Bombay Natural History Society
BSI	Botanical Survey of India
CES	Centre for Ecological Sciences
CSIR	Council of Scientific and Industrial Research
CTS	Centre for Theoretical Science
DOD	(U.S.) Department of Defense
FAO	Food and Agriculture Organization (UN)
FRI	Forest Research Institute
GEF	Global Environment Facility
GHNP	Great Himalaya National Park
IAS	Indian Administrative Service
IBWL	Indian Board for Wildlife
IBP	International Biological Program
IFS	Indian Forest Service
IISc	Indian Institute of Science
IPT	Indian People's Tribunal
SCB	Society for Conservation Biology
SLOSS	(debate over) Single Large Or Several Small (reserves)
UNDP	United Nations Development Program
UNEP	United Nations Environment Programme
UNESCO	United Nations Educational, Scientific, and Cultural Organization
USFS	USDA Forest Service
USFWS	U.S. Fish and Wildlife Service
WII	Wildlife Institute of India
WWF	World Wildlife Fund
ZSI	Zoological Survey of India

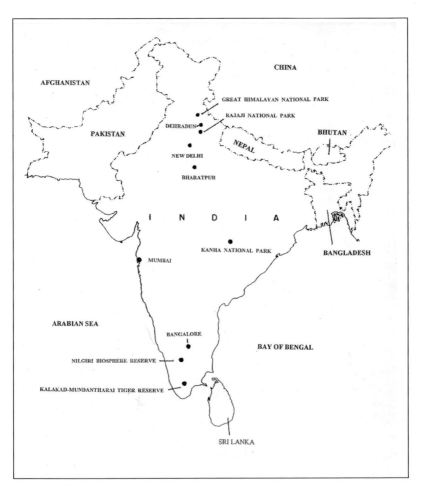

Map 1. Map of India

Tracking Elephants

AS THE SUN ROSE OVER RAJAJI National Park, I followed A. Christy Williams into the Indian jungle in search of Asian elephants and the praxis of Indian ecology. Williams, a researcher at the Wildlife Institute of India (WII), was tracking elephants for his Ph.D. research on the range size, seasonal migration, herd composition, and feeding habits of the highly endangered elephant population in Rajaji.[1] Ramesh, one of Williams's five field assistants, carried the radio antenna that would enable us to find our way through the hills and ravines toward the elephant, and the driver, Govind, prepared for a long rest on the backseat of his jeep—his job was done. For the last hour he had driven us through the forest on dirt tracks as we tried to find the signal from the elephant's radio collar. At each hilltop Govind would stop and Ramesh would hop out of the jeep and unfold what looked like a rooftop TV antenna. Rotating it 360 degrees in short, choppy motions, he paused just long enough between each move to listen for a telltale click from the receiver. Williams had seven elephants radio-collared, with each collar transmitting a different frequency. Five were matriarchs (with their own group of other female and juvenile ele-phants), and the other two were solitary males. Today we were tracking one of the females.

We had finally picked up the radio signal on our fourth stop. As Williams explained, the topography of the park made picking up a signal dicey at times. Rajaji National Park is in the Sivalik Hills, a range formed by glacial runoff from the Himalaya Mountains, some forty kilometers away to the north and east. The Doon Valley runs between the two moun-tain ranges. The rugged Sivalik Hills reach a height of a thousand meters above sea level but drop to about two hundred meters where the Ganges River bisects the park. The glacial runoff is highly erosive and there are numerous steep ravines with small creeks at the bottom. A radio-collared elephant in the bottom of one of these ravines can be invisible to the radio antenna from most points in the park. Conversely, an elephant on top of a high ridge can sound as if it were fifty meters away instead of five kilo-meters. In addition to this topographic variability, the slopes are covered with sal trees (up to forty meters high) and scrubby underbrush.

An elephant standing stock still in the jungle is practically invisible to the eye until you bump into it. Williams has spent entire days searching for an elephant, to no avail. As he said time and again, anyone who has watched TV shows about the African elephant moving majestically across the savannah has no idea what it means to track an Asian elephant in the thickly forested Sivalik Hills—an admittedly large needle in an even larger (800 km^2) haystack.

After another radio reading at the jeep to insure that we started in the right direction, we were off. Our first obstacle was a dry riverbed about twenty-five meters wide. With characteristic understatement, Christy sug-gested that tracking was a trifle more difficult in the monsoon, when be-tween a hundred and fifty and two hundred centimeters of rain fall in just a handful of weeks, washing out the roads, filling the riverbeds, making the jeep useless, and forcing him to travel exclusively on bicycle and foot. The fast-moving waters of the monsoon river had polished the round stones of its bed into smooth and treacherous marbles and bowling balls, ready even when dry to roll under the most carefully placed foot. Williams had snapped a knee ligament on the Rajaji rocks.

We walked and stumbled along the dry riverbed until Williams and Ramesh found a slope slightly less than ninety degrees on the opposite bank and we began our climb, slipping and sliding up a crumbling hill-side, grasping roots and branches to keep from falling. Without pause,

Figure 1. In a small clearing created by elephants in the forests of Rajaji National Park, Christy Williams (in baseball cap) consults with his field assistant as to which direction the radio telemetry signal seems to be coming from.
Cassandra Kasun Lewis, March 1998

Williams and Ramesh floated effortlessly upward while I struggled along. Finally, we stopped to check our bearings, but the unfortunate click of the radio receiver directed us still upward.

After only a couple of minutes, Williams and Ramesh stopped to check the signal once more on a relatively clear hilltop near the top of the range. As Ramesh circled with his antenna, Williams pointed out some huge elephant footprints (called pugmarks by trackers and biologists). There also were some old dried elephant-sized droppings around and several dead trees. Williams explained that this clearing had probably been a resting spot for the elephants. By eating a layer of bark, they had killed the dead trees that surrounded us.

Ramesh and Williams consulted in Hindi. Although Williams is from Tamil Nadu, in the South of India, where he studied Tamil, English, and French in school, he learned Hindi when he came to Rajaji and needed to communicate with forest officials, his field assistants, and the local villagers. Ecologists working in India do not have to cross national borders to do research in exotic locales.

The signal was now coming in very strongly. This meant that Ramesh heard a click when he was facing anywhere within a forty-five-degree arc. It is difficult to imagine how hard it is to differentiate between louder and softer clicks (the loudest click indicating the proper direction of the transmitter), especially when your body is demanding that the click come from the downhill side. Ramesh unhooked the antenna from the wire leading to the receiver and using only the wire still was able to pick up the signal when facing downhill. "The elephant could be very close," Williams whispered, "down in that ravine or somewhere on the opposite ridge. Or she could be on one of those higher ridges."

Following an elephant trail, we slowly picked our way downward. A week earlier, another biologist at the WII had told me that seeing an elephant charge in the jungle could turn the strongest heart to jelly. "It's not just the size or the speed of the elephant but the fact that they come at you in a straight line. They just obliterate any bushes or small trees in their path." When asked what to do if an elephant charged, he had recommended that we run downhill. "Elephants can go down amazingly steep hills, but they can't do it running." The path that we were on was a testament to the elephants' careful downhill movement, as the trail switched back and forth on the gentlest slope possible. Elephants can go down even steeper hills; Williams had seen them go down a hill by sitting on their bottoms and sliding.

We crossed two more ravines before coming to another wide hilltop clearing. We had hiked for several kilometers and for three hours now. For Williams's entire first year of research, he was required to work without radio transmitters and spent many days hiking without sighting an elephant. His project is the latest in twelve years of WII research on the elephants of Rajaji, but the first to use radio transmitters. With funding from the U.S. Fish and Wildlife Service, the National Geographic Society, and the Biodiversity Support Program (a combination of several U.S. groups), Williams was able to get the expensive radio transmitters, the jeep, and the GPS (global positioning satellite) computer and receiver that enabled him to accurately map the elephants' location.

Getting the technology on the elephants was not a simple matter, however. Not only did Williams have to overcome the considerable obstacle of actually putting a collar on the elephants' necks, he also had to get the Indian government's permission to interfere with a protected species. Darting an elephant with a tranquilizer is somewhat risky. Once darted, an elephant usually runs off through the jungle. The scientists have to follow as quickly, but safely, as possible. When the elephant finally collapses, it may fall in the sun (which would overheat it) or in a stream (where it might drown). Though it defies belief for someone who has not been in Rajaji, it can be very difficult to find a sleeping elephant in the jungle. In

1986, when the Rajaji elephant studies were first beginning (still funded at that time by the U.S. Fish and Wildlife Service), an elephant died when it was being tranquilized for radio-collaring, and the government forbade collaring until the WII scientists could prove that the process was safe.[2] Even with funding in place, it took over a decade before the government gave permission for the necessary capture and release of the elephants and to determine that the use of the radio technology would not harm the animals. Such permissions are not always forthcoming.[3]

Even with the technological assistance, tracking elephants was tough. But at the next hilltop we had success—here we found fresh dung. Williams bent down and sifted through a portion of it, noting in his field book what the elephants had been eating—primarily bark, in this sample. Now the elephants did seem to be quite close. Warm dung, unlike radio clicks, tells no lies. Silently, we walked single file down a gentle slope. Here the forest was not as thick. Sunlight hit the ground in many places and the trees were a bit more scattered. At the edge of the deeper forest a small group of sambar (a large deer) saw us and fled quietly into the dark. As we walked, signs of recent elephant feeding were all about us. On nearly every other small tree or bush limbs had been stripped off or left hanging precariously. One large tree had a huge gaping wound on its bark. The elephants, Williams explained, would rub their tusk (if male) or tush (a miniature tusk on females) against the outer bark, until they reached the nutritious inner bark. They would then carefully strip this inner bark, eating the delectable cambium and phloem. Although this group of elephants had left the tree alive, now that the inner bark had been exposed on part of the tree Williams thought that it was likely that other elephants would see the tree and go for the easy food rather than starting the difficult task of exposing the inner bark on a new tree. The days of this tall tree were numbered.

As the clearing turned once more to thicker forest, we began to hear crashing noises. The radio tracking was no longer necessary. We could clearly hear the elephants feeding ahead of us. Williams gestured for me to put my dark blue windbreaker back on—in the dark forest my off-white shirt would stand out to an elephant. Ramesh was the first to spot the elephants. After a few seconds of staring through the woods, I could make out the bulk of one elephant lying in a sunny spot surprisingly near to us. A few seconds later I could see a small calf, just over a meter high, standing nearby, and then an older juvenile a little further to the right, and finally a massive adult munching on branches just behind the sunny spot, partly shielded by a tree. When I focused through my camera we were only nineteen meters away from the nearest elephant, the one resting. The click of the shutter sounded like a firecracker, but the cracking branches were louder.

When Williams finds radio-collared elephants he records their location by means of his GPS receiver, which gives him a precise latitude and longitude reading from satellite triangulation, notes the group's composition, and then records their activity once per minute until they spot or smell him. After quickly jotting some notes in his field book, Williams melted away into the forest to circle around and see how many elephants were in the group. We had been tracking Mallika, the matriarch of a group of six or seven. Her oldest son was on the verge of assuming the solitary life of a male tusker and was only occasionally with the group. Mallika was named on the basis of a resemblance to a classical dancer from the South whose long eyelashes and liquid eyes had impressed Williams in his youth. The group also included two other adult females (one named Madhuri, after a heroine of the Indian cinema), a subadult female (around fifteen years old), a young calf (under two), and a juvenile/subadult about five. Madhuri was the large elephant we could barely see in the shadows, and the subadult female was the resting elephant near the two youngsters. By circling, Williams hoped to locate Mallika and the unnamed adult female.

Suddenly one of the elephants got scent of the intruders and let out a trumpeting squeal. All the elephants looked up, the juveniles crashed through the woods to their mothers, and we ran back through the forest. The crashing continued, but we could no longer see anything. Ramesh signaled us to be quiet. He was preaching to the converted. Williams was still out of our sight and I was worried that some of the crashing noises might be directed at him.

Every student at the Wildlife Institute is told the story of N. R. Nair, a former director of the wildlife training program for the Indian Forest Service (before the WII was actually formed). He and A. J. T. Johnsingh, still on the faculty there, had gone into a forest to photograph elephants. Although a local forest guard had warned that the elephants had been uneasy lately, they proceeded into the forest. When Nair crept up for a picture (Johnsingh hung back), an elephant picked up his scent and charged. Johnsingh was able to get up a tree, but Nair was too close. He tried to use a large tree as a barrier, but the elephant reached behind the tree with its trunk and threw him through the air. He died instantly when he hit a tree. (A domestic elephant can pick up and move one-ton logs.)

Elephants at Rajaji had killed at least six people in the last two years, all people living in or near the park. Elephants, particularly solitary males, often indulge in crop raiding in surrounding agricultural areas. This makes good nutritional sense. An elephant gets many more calories per hour when feeding in a wheat or sugarcane field as opposed to the forest. For males, this translates into added bulk and an advantage in breeding competitiveness.[4] An elephant can eat quite a bit in a night, and farmers are not happy with this. Beyond the obvious conflicts resulting from crop raiding, several towns and villages also surround Rajaji. There are many small settlements and the sacred city of Haridwar along the Ganges. As the river cuts the park in two, elephants often come into contact with these settlements, though they do not venture into Haridwar (it is too large). Every twelve years a religious ceremony (Kumbh Mela) attracts millions of pilgrims to Haridwar, where the Ganges "officially" descends from the mountains into the plains. Not including pilgrims and Haridwar residents, there are still over two hundred thousand people living within five kilometers of Rajaji National Park.[5]

Within a few seconds the elephants we had been tracking calmed down and the crashing subsided. Soon thereafter, Williams reappeared, completely unruffled. The trumpeting squeal, he announced, had been the youngest juvenile. The adult elephants were never really bothered—we were still a safe distance away. Still, as they were alerted to our presence, we should leave.

As we walked out, Williams told of his encounters with elephants. He explained that the animals were really quite safe. Only one time in his two years did he really fear for his life. As if to prove his point, we emerged from the forest onto a small settlement of Gujjars that shared the elephants' territory. The Gujjars are nomadic pastoralists who moved into the (then unsettled and unprotected) Rajaji area about a hundred years ago. They grow no crops and eat no meat. Their entire income is derived from selling buffalo milk to people in larger villages, and they then buy vegetables and other staples. Formerly, the Gujjars would spend the winters in Rajaji and then herd their buffalo into the Himalayas for the summer and monsoon season. Now, however, most of the Gujjars stay in the park year round, as their migratory routes are cut off due to development.[6] They live in small thatch dwellings, clustered together behind a fence. As we approached the village of about ten homes, we noticed a single strand of barbed wire, just over my head, about ten meters before the main fence. One resident explained that it is not to keep the elephants out (it would not work) but to provide the villagers with an early warning—a purpose also served by their beautiful Tibetan mastiffs. An elephant in a Gujjar village several years ago had pulled off part of a roof (probably for the roughage), which fell into a cooking fire and ultimately burned down that home and others. After the fire the villagers had put up the barbed wire. That was the only problem these villagers had experienced with elephants, even though they were only two kilometers from the troop of elephants we had just observed.

There are conflicts between the Gujjars and the elephants, but those are of another sort—conflicts which occur in the arena of public policy and opinion, not in the fields or forest. Although Rajaji has been listed as a national park by the state government of Uttar Pradesh since 1983, it has not been nationally approved as such. This is largely because humans live within the park's boundaries, and that is forbidden by the Wildlife (Protection) Act of the government of India. The law also prohibits the use of national parks by surrounding villagers for fodder, grazing, or nontimber forest products. This law led to the forced relocation of thousands of people, and even complete villages, that were in national parks all over India when the law was first passed by the Indian parliament in 1972. The first relocations had actually occurred in 1968, when villagers had been relocated from Kanha National Park in central India by the Indian Administrative Service officer who later was responsible for most of the writing of the Wildlife (Protection) Act.[7] Unlike the United States, where the national park model developed, India did not have many open spaces with few to no human inhabitants (even in the United States, these "open" spaces were often a product of smallpox and the U.S. Cavalry).[8] Almost all Indian national parks had people living in them when the parks were first established, both rural villages and tribal communities. India's vast population includes approximately fifty-two million tribal people, a number roughly equal to the population of France, and many have traditionally lived in forests.[9] Not surprisingly, many villagers and tribal people resisted being moved from their homes—with variable success, depending on the IAS officers in their district.

Most national parks still do not meet the law's criteria of no human habitation and no human use of resources. The Indian Forest Service is engaged in a struggle to bring these national parks into conformity with the law, a struggle locating them between nongovernmental organizations (NGOs) opposing the enforcement of the law and environmentalists insisting on its immediate implementation. The government of India does not wish to magnify the scope of the debates already occurring in and

Figure 2. This Gujjar village is in the heart of Rajaji National Park. Many scientists and environmentalists want the Gujjars and their livestock to be moved out of the park. As Christy Williams and I met with a villager and a local forest officer, the Gujjars made quite a show of pouring feed for their cattle as if to prove that the cattle do not harm the forest at all, but eat right inside the village boundaries. *Cassandra Kasun Lewis, March 1998*

around those parks by adding yet another "park with people" to its lists.[10] The government is formally recognizing no new parks that do not already conform to the Wildlife (Protection) Act. Thus, the forest officers at Rajaji are attempting to remove the Gujjars from the park (as well as keep surrounding villages from using park grasses for weaving) so that its final notification as a national park can take place.

The Gujjars are fighting their removal, and they have the support of a number of NGOs that are part of a loose alliance sometimes referred to as the "pro-people" movement.[11] The activists also work with villagers just outside Rajaji, trying to help them to get access to nontimber forest resources. These pro-people activists are motivated by the belief that people are not inherently harmful to ecosystems—in India ecosystems have evolved within the context of human use—and that "the solution of the problem [of habitat degradation] does not lie in the eviction of Gujjars but in using the Gujjars in the reconstruction of the forests for a meaningful exploitation of these resources."[12] Either the law should be changed or the park should never be officially notified, they say again and again to any official or newspaper that will listen. Rajaji, only six hours by train from the national capital, New Delhi, has emerged as a very visible test case, both for those who argue that the Gujjars cause no harm to the forest and should be allowed to stay and for those who argue that humans inherently interfere with the natural workings of an ecosystem and must be excluded from nature preserves.

In 1994 the Indian People's Tribunal (IPT) on Environment and Human Rights came to Rajaji to investigate the debates surrounding the park. The IPT was founded to "encourage victim communities to fight for their rights," "highlight the imperatives of equity and human dignity in the search for true development," and "highlight the environmental and human rights abuses being perpetrated on communities and individuals by the ruling elite in pursuance of unsustainable 'development' objectives."[13] This tribunal met with the Gujjars, NGOs, forest officers, and scientists. In a carefully documented report prepared by a former supreme court justice, P. S. Poti, the tribunal suggested that the Gujjars not be required to leave the park, but if they ever wished to move, every effort should be made to help them. The park should be notified, they felt—its biodiversity value was great—but forest policy should not be made without consideration of the rights of the Gujjars and other villagers dependent on the forests for their livelihood. In short, the law should be changed so that the Gujjars, the forest, and the animals would all be treated fairly.

In opposition to this position are many forest officers and wildlife biologists. Faculty members at the Wildlife Institute of India (only twenty kilometers from the park) like A. J. T. Johnsingh, Williams's advisor, have long advocated the official notification of Rajaji National Park and the removal of the Gujjars. As the first director of the WII wrote in a newspaper article about Rajaji, "There must be a well looked after network of parks and sanctuaries from where decimating human pressures are kept at bay."[14] There is no question that for these scientists, the Gujjars' presence is "decimating" the park. As the Gujjars no longer migrate out of Rajaji in the summer, and as their population increases (currently estimated at two to four thousand families, with corresponding numbers of livestock, living within park boundaries), they are putting unsustainable pressures on the flora of the park and causing habitat degradation. They are hurting the Rajaji forests and competing with the elephants and other animals for increasingly scarce forest space and fodder. In the West we are much more aware of the plight of the African elephant, but there are ten times as many African elephants as Asian elephants. Rajaji has one of only two viable Indian elephant populations in the north of the country, and for the sake of the species, Johnsingh claims, this Rajaji population must be saved. Further, by saving the Rajaji elephants, the biodiversity of the entire park would be preserved as a representative ecosystem of the terrai (the belt of land adjoining the Himalayas).[15]

Johnsingh realizes that forced relocation is not a viable option, but he feels that an attractive resettlement package that the Gujjars help devise "would make the scheme workable and save Rajaji." The key, though, is that the Gujjars must go or the park will die: "The ecological disaster looms large and the lifestyles of the people already show signs of advanced unsustainability." The Gujjars respond that the problem is not too many Gujjars but too many elephants. Removing the Gujjars will not solve the problem, they insist. The elephant population is too large for the forest, and the elephants are eating the trees away.[16]

Christy Williams's research will directly address some of these conflicts. His study of migration patterns will suggest the reserve area needed for one elephant or a population. His study of feeding habits will indicate exactly how much, and what, elephants eat in the forest and just how damaging elephant feeding is to a forest. His observation of the different elephant groups and their compositions will help suggest the population dynamics of the elephants in the park and their long-term viability. His surveys will indicate how many elephants currently live in the park, and the population's age structure. His thesis will be ammunition in the Rajaji debates, whether he wishes it or not.

When I asked Williams what he thought about the Gujjar debate, he paused for a while before answering. Although he is a Christian and from a "privileged" caste, Williams strongly defends the reservation system, which is like U.S. affirmative action but much more omnipresent and wide-ranging, giving benefits to a majority of the Indian population at the expense of the numerically smaller but socially much more powerful castes. Williams clearly had a good rapport with the Gujjars and saw them as people, not just as faceless threats to the forests. He believes conservation biology is useless when it comes to actually saving biodiversity. The only way to save wild areas in India—species or landscapes—is through social planning. He stresses the need for curbing population growth and for improving people's material lives, so that they don't need to go into the forest for fuel wood. He spoke of the need to educate people; hardly any of the Gujjars go to school. These things, he said, were the keys to preserving the environment. Conservation biology could provide information about nature, but in the face of increasing human pressure it was not this information that was needed. Pressed further, though, about whether or not he felt the Gujjars should be displaced, or whether Rajaji should be a national park, Christy quietly said, "The elephants have no voice. As a biologist, when it comes to public debates, I feel that I have to be a voice for the elephants. Everyone else has a voice, and the elephants have only the biologists."

CHAPTER ONE

Tracking Ideas

Transnational Science

IN OFFERING TO BE A VOICE for the elephants, Christy Williams is using a directly parliamentarian metaphor to describe his role as a conservation biologist. In so doing, he is indicating his awareness of just how political the science of conservation is. The science of ecology, its subfield conservation biology, and its attendant variety of environmental activism, became a powerful global political force in the second half of the twentieth century. By the start of the twenty-first century, the idea of international treaties to save specific species or entire ecosystems no longer seems novel. Using scientific theories and data as their basis, governments, organizations, and individuals have pushed to literally shape the surface of the globe into a more conservation-friendly place. As the nomadic Gujjar people would point out, however, one person's elephant reserve is another's traditional home. So whose version of conservation is being promoted by the science of ecology? An increasing number of scholars and activists would answer this question with the claim that ecology—specifically conservation biology—and the peopleless national parks that ecologists often advocate are U.S. cultural exports to India and the rest of the world.

Within India there is no question that U.S. ecology—represented by scientists, theories, training, publications, and funding—is omnipresent. Almost every practicing ecologist in India cites American George Schaller's *The Deer and the Tiger* as the most significant early work on Indian ecology.[1] Indo-U.S. collaborative conferences on ecological science held in the early 1980s provided many of the current ecologists in India with contacts for assistance and funding when they were still quite early in their careers—even providing some young ecologists with leads for graduate study in the United States.[2] Journals such as *Conservation Biology, Ecology, Oikos,* and *American Naturalist* are considered (along with the British journal *Nature*) to be the most prestigious of publications and must reads for practicing Indian ecologists.[3] The founder of the Centre for Ecological Sciences at the Indian Institute of Science received his Ph.D. in ecology at Harvard, and when I was there in 1998 the only other ecologist at that institute not trained by him received her Ph.D. in ecology at the University of Miami. Johnsingh, Williams's advisor and the senior biologist at the WII, was trained in field techniques in Virginia by the Smithsonian Institute after becoming interested in ecological research while working as an assistant for Michael Fox's study on the Indian dhole (a wild dog) in southern India. Among the staff at the WII it is assumed that all faculty will eventually have a study trip to the United States—most already have. The funding for large-scale research at the institute often comes from the U.S. Fish and Wildlife Service and the USDA Forest Service (both of which require that projects have U.S. collaborators). The Centre for Ecological Sciences is supported by grants from the MacArthur Foundation, the Ford Foundation, and the Biodiversity Conservation Network, among other U.S. agencies. Almost all the tracking equipment and many of the jeeps used by researchers in India are paid for with grants from the United States. When ecologists in India describe their research, in interviews or in writing, the words and theories they use to give meaning to their data are almost identical to those used by ecologists from the United States, scientists who first described island biogeography, minimum viable populations, and some competition models. More broadly, U.S. ecologists dominate international arguments about the science of ecology, edit many of its journals (thus deciding what gets published and who is considered an expert), and serve on many of the grant committees that award funding.

Williams's research on elephants is intertwined with U.S. interests. U.S. grants have supplied him with state-of-the-art U.S. technology and have paid for several research assistants. After our visit, Williams spent time at Arizona State University, working with his project's U.S. collaborator (a condition of one of his grants) and writing up his results. Williams's advisor at the WII received his training in radio telemetry from the Smithsonian Institution. It is clear that Williams's research is important. Whatever is decided at Rajaji National Park will directly affect thousands of people and will in all likelihood be used as a precedent in many other similar debates throughout India. Ecologists wield power in the contemporary world. So, when Williams chooses to be the voice of the elephants, is he also speaking for U.S. ecologists and a non-Indian set of cultural values and concerns? When scientific ideas cross the globe, not to mention grant money, jeeps, and radio transmitters, do they carry culture with them?

Three Models for How Ideas Circulate throughout the Globe

For many years now scholars in the social sciences and the humanities have been grappling with the question of how ideas, in my case scientific ideas, spread and take root throughout the globe. Clearly the spread of scientific ideas is not a new phenomenon, as even the most cursory glance at the history of algebra, Galenic medicine, or gunpowder would show. Without a doubt, though, these questions have quite a different feel in a time when the East India Company is no more, but economic disparities between nation-states are still stunningly vast and e-mail can cross the globe in seconds. Of the many ways that scholars have tried to explain the movement of scientific knowledge in an increasingly interconnected but still diverse globe, three have seemed particularly influential: models of scientific diffusion, of cultural imperialism, and of globalization. Although none of these models are completely satisfactory by themselves, each offers some insight into how we might understand Williams and his elephant studies—and by implication, the relationship between U.S. and Indian ecology.

The words used to name each model tell us much about each theory. Most of us learned about diffusion in our grade school science lessons. Diffusion is what happens when red dye is dropped into a corner of a swimming pool and all the water eventually becomes pink. Other than the introduction of the dye, diffusion occurs without human agency. It is natural. It is inevitable (imagine trying to keep one corner of the pool free of dye). Models for scientific diffusion propose a spread of science in which knowledge gradually seeps from the scientific center to the periphery in an inevitable one-way flow requiring little in the way of human agency or political intervention. All that is needed is for the more scientifically developed state to make the needed connection with a colony or scientifically dependent state and the flow would start. Obviously, European imperialism provided the classic case for the start of this diffusion. Once the imperial state had left, the communication and exchange network established during the colonial period was in place to keep the diffusion going. The dependent (Third World) scientists could continue to benefit from the scientists and science of the center until some time in the future when scientific expertise and resources had balanced out.[4]

For proponents of this model, it answers nicely the question of why, decades after political independence, the scientists of so many nations still are oriented toward a few scientific centers in terms of training, publications, funding, and project selection, "long after independent political institutions have been erected and long after legions of competent scientists have been put in place."[5] This is the same question we might ask of Williams's research. Why is he, along with other Indian ecologists, so tied to institutions and scientists far from India? Diffusionists would answer that it will take a while—maybe several more decades—for Indian institutions to develop strong research traditions, and for Indian scientists to develop as mentors and teachers for young scientists, and for the Indian educational system to produce young scholars properly acculturated to the scientific tradition and methods. It took Britain a century to get from Newton to Darwin, after all.

Though this idea is now commonly decried by academics as problematic due to its whiggish notions of development, progress, and science, scientific diffusion provided a framework within which people could begin to grapple with this question of science in former colonies that still looked abroad to a distant center. One of the significant formal statements of this idea was an article published in the journal *Science* in 1967; it is not surprising that many scientists view the spread of Western science as a fairly benign and desired phenomenon. Many people share the belief that the development of scientific traditions in developing nations is a good thing and, further, that scientific knowledge diffuses throughout the world in a fairly apolitical fashion. The popular (modern) idea of an "invisible college" of scientists throughout the world, connected by modern communications and selected by their interest in a given scientific topic, similarly affirms a democratic system of knowledge production where all comers are welcomed once they have gotten up to speed. Of course, getting up to speed is sometimes easier said than done, and being "trained" is not as apparent as it might seem.

Diffusionist models were more often cited as cannon fodder than canonical texts over the last twenty years as historians and sociologists increasingly problematized diffusionist models of benign science. One of the most obvious problems of this model is its portrayal of the flow of knowledge and information as a one-way process. Studies of the reception of imperial science have shown that the exchange of knowledge was anything but one-way, and that peripheral scientists and settings had a dramatic influence on science both in the imperial center as well as in the local colony. Richard Grove persuasively argues that knowledge was created at the interface between the center and periphery and that often knowledge produced in the colonies flowed back to the center in the reverse of the diffusionist model.[7] The imperial center did not have a monopoly on new scientific knowledge.

There is another obvious problem with diffusionist understandings of science in the developing world. Indian ecology, as seen with Williams, is no longer oriented toward Britain, India's former colonizer, but toward the United States. This change in orientation is not unique to India. Following World War II, in many scientific disciplines the United States replaced Britain, and other former colonial powers, as the scientific center of much of the newly postcolonial world. Australian and U.S. historians of science writing for an edited collection of essays confronting the question of international and national sciences found that "by the Second World War the U.S. had . . . attained considerable influence within the Australian scientific community. This trend was already clear as Americans extended their biological field research efforts into the antipodes . . . bolstered by the financial resources of American universities and foundations."[8]

One of these historians suggested that "the scientific center is not just people or ideas and certainly not just a nation, a region, a continent, or any geographical place. It is rather a network of institutions, an infrastructure that links knowledge to power through the control of the process of knowledge construction and knowledge communication."[9] It is not actual political colonialism that leads to colonial, or dependent, science in the postcolonial world. Dependent science results from scientific centers having more access to funding, control of journals and libraries, professional societies, and of course the power to advocate certain ideas and people versus competing models and scientists. With a sudden economic or military

Figure 3. Inside Keoladeo Ghana National Park, Bharatpur, the author stops near a sign mixing humor and a bit of self-conscious Orientalism. The sign targets the thousands of foreign tourists who walk and bike the park's main viewing paths looking for the famous cranes.
Cassandra Kasun Lewis, December 1997

boom, today's periphery can become tomorrow's center, with the attendant privilege of defining the terms of scientific debates.

The impact of power on the spread of science is central to the second of our models: cultural imperialism. Cultural imperialism is most often used to address the spread of popular culture or the media (the spread of blue jeans, MTV, or McDonald's), but its rubric has also been used to explain the spread of scientific ideas and institutions. Often, cultural imperialism models focus on the impact of Western power (often termed as control of money, media, and global political structures like the United Nations) in spreading Western values, at a profit. Decades after the postcolonial nations achieved their independence, Western culture is doing what all its armies could not, they claim: turning postcolonial societies into dependent ones—this newest spread is "more insidious and more effective in cementing the dependency of the post-colonial periphery than the fiscal crudities of earlier decades."[10] After all—who would not want to buy the world a Coke and a smile? This form of imperialism relies not just on state power, but also on winning the hearts and minds of its subjects through advertising and Western-oriented education.

While models of diffusion accepted the basic premise that the practice of ecology could be translated between different regions of the world, scholars proposing models of cultural imperialism attacked the claim that science transcended culture or nationality. This argument was based on recent work within the sociology of science suggesting that science is a product of specific cultures, specific agendas, and specific people.[11] From this assumption, it was a short leap for historians and social scientists to argue that Western science was a product of European culture. There is no such thing as international science, just as there is no international culture. Instead, there is Western science, or even more specifically, German or American or British science, which may claim to be international but is in fact imperial in its spread. The attempt to develop scientific institutions on a Western model within a non-Western nation would therefore be difficult because of the cultural incompatibility of a Western cultural product in a non-Western context. It should not be surprising, according to this view, that scientists in the developing world were oriented toward the West as they were subject to what amounted to a form of intellectual imperialism. To do (Western) science meant to be oriented to Western culture.

Cultural imperialism can be used to explain Williams's elephant study. Ramachandra Guha, an anthropologist and historian, has made the claim that ecology, as it is practiced in India, represents the inappropriate imposition of American environmental ideas upon the Indian people and landscape, via "conservation imperialism."[12] Guha subscribes to the idea that science is a social product, and he sees the science of ecology as being predicated largely upon U.S. cultural attitudes toward nature.[13] U.S. ecologists go into nature and see what they are culturally prepared to observe. This is to be expected—culture cannot be escaped entirely—and Guha does not criticize this tendency within U.S. ecology.

What concerns Guha is the export of this science to other places that do not share the same cultural or physical characteristics. Thus, for instance, how appropriate is a U.S. ecology predicated upon humanless study sites in Wyoming or Alaska to a densely and anciently populated land like India? Not at all, says Guha. To return to Williams and Rajaji, in the conflict between Gujjars and elephants a science developed in the New World is not appropriate in the Old. From Guha's perspective, Williams's scientific agenda is dominated by U.S. cultural conceptions of nature written into U.S. ecological models and funding decisions. This science would be appropriate for debates about grizzly bears in North America, perhaps, but not elephants in North India. Guha calls instead for a local Indian ecology, derived from local customs and knowledge and oriented toward the solution of local problems. This is not a popular idea among ecologists—one article by Guha was circulated among some Indian foresters and ecologists under the e-mail subject heading "crap.doc." In response to Guha, scientists in both India and the United States argue that ecology *does* produce site-specific knowledge. Williams may use U.S. methods, U.S. funding, U.S. collaboration, and U.S. models, but his study will generate Rajaji-specific results. His data will reflect the actions of the elephants at Rajaji—not foreign models. How feasible is such a balancing act, though—is there a point when the methods and models of the study do influence the final shape of the results? Yes, cultural imperialists would claim.

Shiv Visvanathan, a fellow at India's elite Centre for the Study of Developing Societies, makes a more wide-ranging critique of the practice of the science of ecology in India and throughout the globe. Visvanathan is a vocal critic of Western-style development policies that he sees as ultimately destructive. He believes that Western science provides the foundation for development: "underlying modernity was the social contract between the nation-state and modern western science and both were engaged in a soap opera called development. The process was becoming life-threatening and even genocidal."[14] Western-style development is inextricably linked with the modernist belief in the possibility of managing the natural world, and dependent upon modernist sciences used for massive hydroelectricity projects or atomic energy.[15] Ecology is implicated in Visvanathan's critique insofar as ecologists make attempts to manage nature or to change the local (Indian) landscape in the interests of an international environmental agenda. Instead developing countries like India need a local (not international) ecology that would make recommendations within the context of Indian culture and folkways.[16] Removing thousands of people from their homes to create a peopleless national park is not appropriate in a densely populated land like India. Like Guha, Visvanathan laments the fact that so many Indian scientists and government officials have bought into what he sees as strictly Western science and development. When I spoke with Visvanathan about my project in his office in Delhi, he encouraged me to change my plan of study. Rather than investigating Indian ecologists who work in the Western tradition, I should go live in a Rajasthani village, he suggested. There I could get a sense of what Indian ecology really was.

As with diffusionist models, cultural imperialism models overemphasize the power and singleness of purpose of the West and severely limit the agency of non-Western actors. In trying to move past a vague notion of influence and take into account the very real relations of power between the developed and developing world, cultural imperialism scholars can run the risk of overstating their case.[17] These explanations often posit a passive Third World audience for the cultural exports of the West, and they often claim that an authentic Third World culture is being corrupted by an inauthentic First World culture. There is little room left for agency by actors in the developing world in the "strong" version of cultural imperialism. This is a major flaw. It seems clear, for instance, that Williams chooses to study (and like) elephants at least partly through his own volition, not solely because of his interpolation by the hegemonizing power of U.S. ecologists.

In response to models of cultural imperialism, some scholars have begun to argue that the spread of knowledge can best be understood through models of globalization, our third and final model for how ideas spread across the globe.[18] Globalization assumes that the importance of geography as a determinant of "social practices and preferences" is quickly diminishing; culture is becoming less a function of locality than of preference in an increasingly interconnected world. People are not hegemonized, globalization advocates might argue, they are merely choosing (Western) alternatives not available to them previously. There are many competing models of globalization at the moment, from consumerist justifications of the McDonaldization of the world (people buy something because they like it), to analyses focusing on the importance of the electronic media in creating a global village, to critiques that globalization is just a favorable term to describe the insidious Americanization of the world.[19]

Arjun Appadurai suggests what is for me the most compelling vision of globalization in his mixture of locality and global processes. He argues that the nation-state is breaking down as a useful referent in the face of mass migrations and the electronic media. Nation-states and their state policies and politics are not useful ways to discuss or analyze the movement of ideas, of people, or of information, he argues. As Appadurai writes with regard to media, "the United States is no longer the puppeteer of a world system of images but is only one node of a complex transnational construction of imaginary landscapes." When a French family watches *Dallas,* what they see (or interpret the show's meaning to be) is different than what is seen by a similar family living in Texas. Local people appropriate foreign images, ideas, or items and use them in a locally specific fashion. Scholars interested in the movement of knowledge should study "global cultural flows" of ideoscapes, not one-way impositions of knowledge from more to less powerful states, Appadurai argues. The strength of this approach is its obvious respect for local differences, reactions, and resistances, and its awareness of the affect of the burgeoning electronic media operating outside state controls, elements

Figure 4. Inside Great Himalaya National Park, two days' hike from the nearest road, these local villagers are looking for mushrooms to harvest and sell to a Dutch multinational.
Photograph by author, April 1998

that are all too frequently missing from more power-laden approaches. In
this analysis, Williams's project might be understood as one node in an international network of ecological experiments, all interlinked and benefit-
ing from their respective experiences and insights. Appadurai acknowledges
the power of local settings in determining people's worldview, but also
draws attention to ways in which the local is increasingly global. For instance, U.S. ecologists might read Williams's results and change the way they
conceptualize their own science on the basis of his findings and insights, insights that might result from his very specific local Indian context. At the
same time, Williams will probably couch his insights in the language (statistical, theoretical, and grammatical) of the United States. This is a much more
subtle interpretation of how ideas move about and are appropriated by different actors for different purposes than cultural imperialism models or simple diffusion models. My primary concern with this model, however, is that
it downplays the role of power relations in what knowledge is accepted and
codified, versus rejected or marginalized. There were rewards for Williams
choosing to conduct his elephant research as he did (funding, potential publications, recognition), just as there would have been penalties had he chosen otherwise. A historian reviewing recent trends in the recent history of

हमें भी जीने का अधिकार है !
WE ALSO HAVE RIGHT TO LIVE.
जी, एच, एन, पी, शमशी

Figure 5. This monal pheasant and slogan, painted on the side of a building in Kulu District, Himachal Pradesh, in the heart of the Himalayan Mountains, is an example of the many attempts by Indian conservationists to develop a First World–style environmental consciousness in rural India.
Cassandra Kasun Lewis, April 1998

science acknowledged the appeal of recognizing local influence in international science but still wondered, "[W]ho sets the research agendas of international science? The implication is that everyone . . . does—but do they do so equally?"[20]

Following Trails, Telling Tales

When I attempt to understand Christy Williams's research—its origins, its methods, and its implications—none of these three models are completely satisfactory. Williams is working thousands of kilometers and oceans away from the publishing houses of the journals to which he would like to submit his work. In very tangible terms, he is on the periphery. But his peripheral project is better funded by U.S. institutions than some projects working in the center itself, on less charismatic species and ecosystems (like insects in Midwestern prairies). It is clear that when Williams claims to be the voice for the elephants, he is more than just a mouthpiece for the U.S. imperial ecologists. He is deeply interested in his topic and brings personal motivations to his work that predate his training as an ecologist or his receipt of U.S. funds. Similarly, it is also clear that the process whereby some scientific projects are funded and supported, while others are allowed to languish, is an exercise in the power to control the boundaries of ecology. For Williams's project, primarily U.S. ecologists made those decisions. For good or ill, U.S. funding agencies have judged his work as "favorable," and his project does conform to the highest standards of U.S. ecological practice. Is that a mark of the current level of scientific diffusion in India, of hegemonization and corruption, or of the globalization of ecological agendas and practices? Who benefits from this ecology and who does not? I have used Williams as my example here, but I could have as easily chosen any number of other projects. Across India, similar ecological research projects bring up similar questions: What did it mean to practice the science of ecology in India in the latter half of the twentieth century? How is Indian ecology linked to U.S. scientists and agendas?

These debates are what drew me to go to the jungle with Williams, as well as the figurative jungles of U.S. and Indian ecology and conservation biology. I went searching not for elephants but for an idea of how ecological theories and practices crossed the globe and of the impact of U.S. science and scientists on the practice of Indian ecology since Indian independence. That search led me from the lush rain forests of the southern tip of India to the breathtaking peaks of the Himalayas, from Delhi to Dehradun and from Mumbai to Bangalore, and from the most recent decisions of the government of India to the height of the Indian nationalist movement in the last days of the British Raj.[21] The story I found is more intriguing than I could have guessed—there are patriots and xenophobes, imperialists and suspected spies, birds carrying diseases and crocodilians carrying radios; there are stories of people collaborating across barriers, but also of human destruction—both of animals and other people. So let us begin by turning to the colonial city of Bombay and the first stirrings of an independent India and an independent Indian ecology.

CHAPTER TWO

The Gateway to India

Sálim Ali, S. Dillon Ripley, and the Introduction of the New Ecology to India

> The day that we approached Bombay, it was of course very early in the morning, and there was this distant hint of light in the sky through the mist and the dust which lay over the mainland as we approached it, because we were coming from the West towards the East. . . . We went of course to the Taj Mahal Hotel, which is the one and only hotel in Bombay, and the most romantic, straight out of Kipling hotel that you could possibly imagine—a Victorian, gothic structure right on the mole, the edge of the harbor right near the Durbar Gate, the Gateway to India, a tremendous Arch of Triumph–like structure. . . . All the magic of the East of Kipling came over us like a tremendous sort of spell. We were enraptured by Bombay, and we were enraptured by our stay there.
>
> —*Dillon Ripley*

THE GATEWAY OF INDIA is one of the most famous landmarks in Mumbai (formerly Bombay), duly commemorated on postcards and travel books. The tall "Arch of Triumph" of yellow basalt stands in a park, facing the Arabian Sea. The arch, built to honor the 1911 visit of Britain's George V, was completed in 1924. Unlike similar but hollow boasts, Mumbai truly has served as the gateway to India for the last hundred and fifty years, both during the British Raj and after independence (in 1947), and is now considered India's capital of commerce. Before air travel, many Westerners entered India through Bombay after sailing across the Arabian Sea, and now the Mumbai and Delhi airports compete as the most popular entry point for international travelers.

A ten-minute walk from the Gateway of India (with its statues and snake charmers) into the heart of downtown Mumbai (with its Victorian-looking banks and double-decker buses) brings you to the Prince of Wales Museum. Tucked next to the museum, in a lot carved out of the museum's original grounds, is Hornbill House, the headquarters of the Bombay

Natural History Society (BNHS).[1] Founded in 1883 by six British and two Indian naturalists, the BNHS has since independence become one of the premier conservation and research organizations in Asia. It is a non-governmental organization, and although it has an international membership, its focus is directly upon the natural history of South Asia, particularly India. The *Journal of the Bombay Natural History Society* (*JBNHS*), is the leading conservation biology journal in India, if not all of the developing world, and its collection of over thirty thousand bird skins (stuffed birds), particularly those from South Asia, is nearly exhaustive, with over 30,000 specimens. The BNHS has played a leading role in environmental advocacy in India, particularly in the establishment of many national parks and the Indian chapter of the World Wildlife Fund (WWF), in the writing and passage of India's 1972 Wildlife (Protection) Act, and in the development of almost all the senior wildlife ecologists currently working in India. Although the BNHS is less omnipresent in contemporary Indian ecological and environmental debates, this is at least in part due to its own success in increasing interest in these topics within India over the last fifty years and its support for the corresponding increase in environmental NGOs (like the WWF), academic programs in wildlife biology, and Indian biologists trained to pursue ecological research. Perhaps more than a gateway, the BNHS has been a colossus astride Indian conservation–oriented ecology over the last fifty years.[2]

The BNHS has also been a gateway for Indian science, though, serving as the point of entry for specifically Western ecological ideas and scientists for over a hundred years. The staff of the BNHS have worked with and supported scientists from throughout the world in their work on the subcontinent. When affiliations were needed to secure research visas to work in India, the BNHS provided them. When field assistants or local contacts were needed, the BNHS helped provide them. And when the research was done, the *JBNHS* brought their theories and reports to an Indian audience.[3] In its early years, the BNHS was understandably linked closely with the British Museum, and most of the naturalists who worked under its auspices were either British scientists or British expatriates—tea planters, members of the India Medical Service, or retired military officers. These men, for the most part, ran the society.[4] Following independence in 1947, however, control of the BNHS shifted suddenly to Indian hands, and the international orientation of the society's leaders also shifted to the United States as a source of collaborating scientists, funding, training, and expertise. Simultaneously, Sálim Ali, who would become the charismatic leader of the BNHS for most of India's independence, explicitly advocated the introduction of the new science of ecology into India, a science he had learned in Germany. Through the example of his own work, as well as his active collaboration with the U.S. ecologist S. Dillon Ripley, Ali hoped to develop a strong ecological tradition in India.

The Birdman of India: Sálim Ali

At the BNHS the imprimatur of Sálim Ali is everywhere visible, and the organization has visibly suffered from his absence since his death, in 1987. He was India's most famous conservationist and ornithologist, and he inspired, encouraged, or trained most of the ecologists currently working in India.[5] His story is the story of the BNHS, from his gradual transition from naturalist to ecologist, from his relative marginalization as an Indian in a largely British organization to his leadership of a postindependence NGO that helped to define how Indians thought about and viewed their natural world, to his growing collaboration with U.S. scientists. In an obituary, Madhav Gadgil, founder of the Centre for Ecological Sciences in Bangalore, remembered Ali primarily as "the man who taught Indians to appreciate, to study at first hand, to treasure, to work towards conserving the rich living heritage of this country."[6] Ali, if not the father of Indian ecology, was at the least a pioneer, bridging pre-independence India with the present, and providing a dramatically different model for the practice of ecology compared to that which preceded him. It is thus worthwhile to focus more closely on the meandering process by which Sálim Ali came to be a field ecologist and on his interests and motivations.[7]

Becoming an Indian Ecologist, before Independence

Sálim Ali's path to becoming India's leading animal ecologist and environmental advocate was not a direct one. Born in 1896 to a large and affluent family of Bombay merchants, Ali had a relatively privileged childhood. Orphaned at the age of three (along with nine older brothers and sisters) he was raised by a childless aunt and uncle, part of an extended family that would continue to sustain him throughout his life. When Ali was around nine or ten, an uncle gave him a repeating Daisy BB gun, and Ali became the terror of all small birds in the area. Living in what were then the outskirts of Bombay, there was always a plentiful supply of game for Ali's exploits. Although in his autobiography he does admit to "elation" if he shot an uncommon bird (5), he quickly goes on to claim that this success was quite rare. He mainly shot the plentiful—and hardly in danger of extinction—house sparrows.[8] A friendly cook would help Ali fry up his catch with a little ghee [clarified butter] and some spices.

Ali began his lifelong habit of taking notes on birds at this time. His goal was not science, however. Instead he hoped to glorify his prowess as a hunter who cunningly learned sparrow behaviors in order to better kill them: "The cock sparrow perched on the nail near the entrance to the hole while the female sat inside on the eggs. I ambushed them from behind a stabled carriage and shot the male. In a very short while the female acquired another male who also sat 'on guard.' . . . I shot this male also, and again in no time the female had yet another male . . . in the next 7 days I shot 8 male sparrows from this perch" (4). These hunting notes were much later published in the *JBNHS* and presented as the earliest writings of a young prodigy—indicative of natural interest and ability in what would be his life's profession. Although no one would mistake these notes for those of a scientist, they do suggest that Ali was attentive to bird behavior.

Ali's sparrow hunting led him to the BNHS in 1908. When young Sálim, then twelve, noticed that one of the sparrows he had shot had an unusual yellow patch on its throat, he showed it to his uncle, the hunter. Sálim was concerned; was this bird safe to eat or was it forbidden by his Islamic dietary restrictions? His uncle agreed that the sparrow was unusual and encouraged Sálim to take it to the honorary secretary (head) of the BNHS, W. S. Millard. Sálim's uncle had joined the BNHS shortly after its founding and was active in its work. He wrote Sálim a letter of introduction and sent him to town.

Ali's only previous contact with Englishmen had been the impersonal annual visit of the school inspector: "They had built around themselves a mythical aura about their nobility and greatness and superior virtues. . . . I remember the feeling of nervousness—almost of fear and trembling—at the prospect of meeting a full-grown sahib face to face with which I entered the quaint old single-storied building through its magnificent solid teakwood portal" (7). Ali opened a door not merely to a building but to what would become his life's passion. As it turned out, Honorary Secretary Millard was a kind man. After identifying Ali's specimen, Millard showed him the BNHS reference cabinets, filled with thousands of different stuffed birds. This was Ali's epiphany (8). Millard gave Ali some books and introduced him to some of the BNHS staff, including an assistant, a young Englishman born in India, S. H. Prater, the man who would suggest twenty years later that Ali write the best-selling *Book of Indian*

Birds.[9] With the invitation of Millard, Ali began to come to the BNHS offices quite often, learning how to skin and stuff birds and other animals, and how to use the collection to identify new specimens. Apparently Millard went out of his way to encourage Ali, and Millard's wife even plied Ali with regular servings of chocolate cake.[10] Ali became a committed amateur naturalist at the BNHS, a hobby solidified by hunting vacations with uncles and brothers. When he entered the portal of the BNHS, Ali entered the domain of British colonial zoology, a science characterized by the collection, study, and classification of the thousands of dead animal specimens found in the BNHS office.

Although Ali entered college thinking he might become a zoologist, focusing particularly on ornithology, he was driven from it after the first year of "struggling hopelessly with logarithms and suchlike evils." Ali was required to pass these prerequisites in order to take any biology classes, but he saw no connection between the formal study of math and science as presented at the university and the type of natural history in which he was interested. Many of the highly flattering accounts of Ali written by his disciples still mention his quick temper and lack of tolerance for what he viewed as inefficiency: "Bad work maddened him—his face would redden, and his head would shake. One look at a shoddy report and he'd fling it aside, storming furiously."[11] It seems likely that Ali had a similar reaction to university requirements with which he did not agree.

When a brother living outside Rangoon and running a tungsten mine suggested he could use some help, Ali immediately dropped out of college. Although Ali does not mention it in his autobiography, he and his brother also were selling timber logged out of the jungle.[12] As a young man, Ali was a full participant in both the culture of hunting (practiced by Indian royalty long before the British arrived, but also a symbol of status in the British imperial state) and in the exploitative resource extraction of twentieth-century colonialism. When Ali traveled to the other side of British India, to Burma, he went to a forested region inhabited primarily by tribal peoples, people as foreign from his background as they were from that of a British civil servant. As a young man, Ali, a somewhat privileged and wealthy Indian, shared more with the British than he did with the Burmese in whose land he worked. As evidenced by his family's friendship with leaders of the Indian independence movement (including Nehru) and personal statements in his autobiography, Ali was an opponent of British colonialism in India. As his work in Burma indicates, however, Ali's anticolonial sentiments did

not automatically translate into him taking a stance against development, industry, or Western science.[13]

Ali was in Burma from 1914 to 1924, with the exception of a year and
a half (1917–18) in which he returned to Bombay for further training for
the business and to be married. Ali's family encouraged him to take a year
course in law and accountancy, which he dutifully completed. At the same
time, however, he joined Prater (with whom he had become good friends)
in the yearlong course in zoology at St. Xavier's College, a Jesuit school
in Bombay. Prater was taking the course in order to become a scientific
member of the BNHS staff, while Ali took the course as a hobby. Both
young men spent their spare time identifying specimens at the BNHS.[14]

When Ali returned to Burma in 1918, newly married and trained in
accounting, surely his brother expected a business partner full of ambition
and dedication to the fledgling business. Ali did return to Burma with a
renewed commitment, but his commitment was to exploring the surrounding jungles. It was a time when he could be the "intrepid explorer
and big-game hunter" he had dreamed about as a teenager (20). Sixty
years later, he remembered fondly "those uncomfortable yet exhilarating
nomadic days in the jungle. . . . [W]hen one has forgotten the mosquitoes
and the leeches, the constant monsoon drenchings all day and the wet,
soggy bedding at night, they even seem romantic" (40–41). Again, the
reminiscences of the upper-class urbanite Ali sound not unlike those of an
old "India hand" remembering his days in the empire to the grandkids in
Manchester. One of Ali's colleagues remembered that when Ali "was sent
into remote jungles to select timber, he'd spent most of his time observing birds and wildlife. Not surprisingly the business collapsed."[15]

When Ali returned to Bombay in 1924 he had decided to be a biologist. There were not many jobs, especially for a biologist with no university degree, and Ali and his wife lived with his family while he searched.
After briefly working for a cousin as a clerk, Ali was hired to be a guide
at the Prince of Wales Museum (for the natural history section), largely as
a result of his friendship with Prater, who was by 1924 the curator of the
BNHS collections.[16] Here he was called upon to educate school groups,
using the BNHS collections. Simultaneously, Ali was continuing his work
on bird observations, and taking weekend hunting trips into the countryside with Tehmina, his wife. His lack of a college degree inhibited his success in finding other employment or in moving up within the BNHS, and
after a few years at the museum Ali went abroad for more training in ecology and ornithology.

From Natural History to Ecology

When Ali went to Europe in 1929, biological study in India was divided into zoology and botany. At the organismal level, both sciences emphasized taxonomy. The study of zoology in practice meant the attempt to understand the form and functions of animals and on the basis of these characteristics to place the animals into taxonomic categories within a larger (Linnaean) framework. An Indian ecologist noted wryly that pre-independence zoological training in India consisted of "the formal study of dead animals."[17] This type of work had been undertaken with great gusto by British civil servants in the 1800s, particularly by members of the India Medical Service, who had acquired a rudimentary training in zoology in medical school. For these scientists, India represented a vast storehouse of new species, never before catalogued and in need of incorporation into the Linnaean classificatory scheme.[18] As Western-style universities were established in India in the 1850s, they naturally corresponded to the orientation of the British scientists at work in India. The colonial government established the Botanical Survey of India in 1890 and the Zoological Survey of India in 1916, with the explicit goal of cataloguing India's natural biological diversity.[19] The BNHS collection on which Ali had spent so many days working in Bombay was also a tool for this aim. By comparing new specimens of birds with the collection of stuffed skins, zoologists could determine where in the grand scheme of nature a new sparrow or hawk might fit.[20]

This is still the case. At Hornbill House, the BNHS building in Mumbai, thousands of species of birds, reeking of mothballs and old feathers, are tucked in cabinet drawers in a massive collection room hidden behind rows of doors. These drawers are among the last haunts of several endangered and extinct species, including the famous, and unfortunately extinct, pink-headed duck, whose pink when I saw it, in spite of the best efforts of the curators and their staff, is getting a bit faded—we will never know exactly how beautiful this bird once was. Indian avian biodiversity was nowhere more tangible to me than in these stacks of cabinets deep in the bowels of Mumbai. Although the BNHS collections include nonbird species, Ali, an ornithologist, so shaped the BNHS and the public's perception of that organization that it has become indelibly associated with birds, despite its important work on many other scientific classes. Saraswathy Unnithan, one of the current curators for the BNHS collection, works with the birds and described how "amateurs . . . can send us a description for identification. Sometimes they will send a bird that was found dead, or they send photographs or slides." In addition to amateurs, she receives requests for assistance from scientists in the field, from museums throughout the world, and from forest officers. Sometimes scientists, often taxonomists, will come to the BNHS to work with the collection.[21]

This work, the working out of the finer details of taxonomy, was all that Indian zoology offered Sálim Ali in the 1920s, when he began with the society. And although he recognized it as important—his collection trips throughout India were often dedicated to the development of this very collection—he wished to engage in a different type of science. In his introduction to *The Book of Indian Birds,* Ali wrote "Our greatest need today is for careful and rational field work on *living* birds in their natural environment, or what is known as Bird Ecology. . . . This is a line of field research that may be commended to workers in India; it will afford infinitely more pleasure and is capable of achieving results of much greater value and usefulness than the mere collecting and labeling of skins." In the same section, Ali noted that "systematic ornithology," the taxonomic work, is a job for experts in museums—clearly not where he wished to work—and that it does not "achieve results" of equal significance to bird ecology studies. He knew that taxonomy was essential—indeed, the precondition for all other work—but he wished to also learn something of Indian birds' behaviors and interactions with the larger world.[22]

Ali was attracted early on to a very different tradition than what was offered by the Linnaean zoologists, to that of the naturalists. Naturalists, the practitioners of natural history, had a long history within India. In the late twenties, before his trip to Europe, Ali wrote a series of papers for the *JBNHS* that described the naturalists among the Mughal emperors.[23] Ali was also very familiar with the British natural history tradition, as practiced by many British civil servants who, while in India, would submit short notes to the *JBNHS* detailing observations of some bird or other animal. A naturalist would certainly be interested in the identification of the animal, but would also note what it was doing, what it was eating, and try to discern some order in its behavior. It was relatively easy to become a naturalist—no scientific training was required to observe—and for this reason "natural history" both in India and elsewhere was often thought of by formally trained scientists as a somewhat questionable activity.[24]

Some scientists within Germany, Britain, and the United States, however, were beginning to develop observational practices that were based in natural history but had more rigorous standards of observation and explanation. One focus was the interactions between plants, animals, and their environments. Ernst Haeckel, called this type of study oecology in 1866. The term remained unused, though, until after U.S. scientists transformed it to *ecology* in the 1890s.[25] Ecology did not necessarily break down the botany/zoology divide. But rather than studying dead specimens or collecting individual plants, these scientists (who often still called themselves botanists or zoologists) began to look at communities of organisms, naturally occurring in the field, or interactions between members of a population of animals. In the United States, the most noted of these early scientists was a botanist, Frederic Clements, who investigated the composition of plant communities. Many of his theories are quite familiar—ecological succession and climax communities, for instance. Within Germany, a group of bird biologists came to the fore, with a focus on the study of animal behavior (called ethology, now considered a subfield of ecology). Birds were singularly well suited for behavior observation in the field, and bird scientists such as Konrad Lorenz and Nikolaas Tinbergen (the Nobel Prize–winning fathers of ethology) led the way.

Whether studying plants, animals, or (rarely) both, ecologists were engaged in a dramatically different type of science. A traditional zoologist or botanist, after walking in the forest, would describe "lists of the species encountered . . . and then speculate about why particular species were present or absent, or why they were represented abundantly or rarely."[26] In contrast, an ecologist might note which species were present or absent, then what they were doing, and wonder how they interacted. Beyond this, ecologists also began to count, measure, and use quantitative methods to describe their observations.[27]

When Sálim Ali decided to travel abroad for training, he sent letters of inquiry to Britain and to Germany. To his regret, he found that "the political atmosphere between India and Britain [in] 1929 . . . was charged with so much bias and bitterness that, as was apparent from the lukewarm response I got from the British Museum, conditions of work in Britain would have been uncongenial."[28] This in spite of the fact that Ali had worked extensively with British biologists in India and knew several members of the British Museum staff.[29] Ali was disappointed—he had strongly preferred to study in Britain, and he had thought that his personal relationships with British scientists would open doors. However, although Ali was unknown to Erwin Stresemann of the Berlin University Zoological, Stresemann did agree to work with him. This was a fortunate turn of events. Stresemann proved to be one of the twentieth century's most famous ornithologists. He was one of those rare scientists who not only excelled in research but also was a superb mentor. He counted among his students, from around the world, some of the twentieth century's outstanding ecologists, such as two of Ali's colleagues while in Berlin, Oskar Heinroth and Ernst Mayr.[30]

Ali thus found himself traveling to Germany to study ornithology with a scientist who happened to be at the center of the move in that field from strict taxonomy to ecology—a far better choice for a year's study than the British Museum. The British Museum in the twenties and thirties was dominated by biologists concerned with taxonomy and evolution (using taxonomy to demonstrate the workings of evolution), and the new move to ecology had not yet made inroads in that setting.[31]

In agreeing to take Ali as a student, Stresemann was not acting purely on charitable urges. He suggested that Ali bring with him to Germany a "collection of Indian Birds to work out under his guidance." A collection of birds from the Indian subcontinent was a valuable commodity for an ornithologist—new species awaited, and a quick glance through Ali's *Book of Indian Birds* reveals that ornithologists were not bashful in assigning their own or colleagues' names as the birds' new common names (Finn's Baya, Loten's Sunbird, Tickell's Flowerpecker, Pallas's Grasshopper Warbler). These birds were even more valuable to Stresemann because, as a German (especially between 1914 and 1945), he had limited access to British India and its animals. For Ali's year in Germany, the BNHS gave him two hundred birds that had been collected by a British member of the Indian civil service at the request of British scientist Claude Ticehurst. When Ticehurst found out that his birds were to be taken and classified by an Indian amateur and a German scientist, he was not pleased and formally protested to the BNHS. The BNHS stood firm, however, and Ali arrived in Germany with the birds.[32]

At first glance it might appear that Ali was being placed in a taxonomic pigeonhole—"Let me help you look for new species"—but in Stresemann's lab he found a community of students and scientists that fundamentally changed the way he thought of and conducted science. Ali went into the field with these new colleagues, he worked with them in labs, and he talked with them over meals. Although Ali could not stay long enough to pursue a degree in Berlin (he spent exactly seven months there),[33] when he returned to India he carried with him the latest ecological theories and techniques and friendships with an astoundingly successful group of young scientists. Both his training and his network of colleagues were state-of-the-art resources. Mayr and Stresemann proved to be particularly constant colleagues and friends. After reading one of Ali's papers in 1936, Mayr wrote, "I hope your work will set a standard for future studies." Stresemann continued to correspond with Ali for the rest of his life, and Ali "considered him my guru to the end."[34]

Upon his return to India in 1930, Ali embarked on a remarkably productive period of fieldwork. Thanks to his German training in the new behavioral ecology, Ali was prepared to undertake a type of research far more sophisticated than that conducted by any of his British peers in India, or in Britain, for that matter. Ali was the first "ecologist," as distinguished from taxonomists and more traditional naturalist-zoologists, to work in India, whether European or Indian. When Dillon Ripley wrote that Ali "introduced the more modern-day concepts of ecology to India . . . [and] helped to revolutionize Indian biology," Ripley did not just mean that Ali was a leader because he was the first Indian to do what Europeans had been doing in India. Ali was truly the first ecologist in India. Ecology was not introduced to India by Westerners, but by the degreeless Ali, by way of his German training.[35]

Ali was still unemployed, though. While he was on leave in Germany, the BNHS had abolished his meager guide job at the Prince of Wales museum, as Ali found when he returned to Bombay.[36] Ali and his wife once again relied on his family for support, living in a family home near the coast. At least one uncle (Ali tells us in his autobiography) saw him as "a shirker and a waster, and [opined] that ornithology was just a cover for my indolence and reluctance to do 'honest and gainful' work."[37] In hindsight, Ali often "admitted that his inability to find a job, 'in spite of his best efforts,' was one of the luckiest things that ever happened to him."[38] Of course, Ali made his own luck in this case, for he could have had any number of nonbiological jobs, had he so desired. During the years after his return from Germany, jobless, Ali conducted and published two superb studies in the *JBNHS,* one in 1931, the other in 1932, that exemplified the sort of behavioral ecology he had imbibed in Berlin.

The first of these studies dealt with the nest building habits of the Baya weaverbird,[39] work that Ali considered the "single event in his life [that] had given him the greatest pleasure."[40] This study of Ali's was (granted, in the words of a relative, fan, and colleague, but also one of the founding leaders of the World Wildlife Fund–India) "a classic field study, enough to insure immortality for any ornithologist" (5). Ali was the first person to do a prolonged study of the Baya weaverbird, a species found throughout India that builds unmistakable large hanging gourd-shaped nests. Ali had noticed a colony of weaverbirds near the house where he and his wife were staying and had decided they would make good study subjects (5). Ali's study was essentially behavioral; he determined that the male bird would build nests that were the basis for mating selection by the female of the species. Female weaverbirds would inspect the half-built nests; if they approved of the nest, the male would finish the nest and the birds would mate. If the female did not approve of the nest, she would fly off to look at another prospect. The male bird would abandon the half-built nest and begin work on another one, in hope of attracting another female. Ali's pioneering work is the basis on which all subsequent studies of these birds have been based.[41]

The second of Ali's two studies in the years immediately after Germany is less well known but is an even better example of the new ecological work he was attempting. It dealt with birds' role in the pollination of flowers. Ali was explicitly introducing a subject that had "hitherto received no attention . . . from workers in India," largely because it was on the boundary between botany and zoology.[42] Topics such as this had not received much attention outside of India, either. Ali made a strong case in this article for the necessity of understanding bird behaviors within the context of the plants that fed them and for understanding plant adaptations on the basis of the birds that cross-pollinated them. This work combined behavioral ecology, evolutionary botany (how plants evolved), and the sort of attention to interactions between different species that was the hallmark of the new ecology approach (though, as pointed out earlier, even most of the new ecologists did not cross the botany/zoology divide).

Although these studies were not published in a Western journal, the work was "way ahead of ecological and behavioral studies not merely for India, but the world over."[43] These studies are thorough and carefully documented examples of the attempt to understand bird behavior in the field, and with the second study, Ali transcended the zoology/botany divide, discussing the relationships among species of birds and flowers. His articles stand out from the other articles in the *JBNHS* like a parakeet amid sparrows. With this published work, Ali began the long process of introducing ecology, a "new" style of natural history, to India. Ali was bringing to the readers of the *JBNHS* (and thus to the offices and homes of India's leading naturalists and ecologists) the theories and results of cutting-edge German ecology, the same German ecology that was moving to the United States in the thirties as Ernst Mayr left Germany for the American Museum of Natural History in that dangerous decade.

The Book of Indian Birds

Sálim Ali closed out his first decade after his return from Germany by finishing the book for which he is best known in India, a field guide entitled *The Book of Indian Birds*. The bible for amateur birders, it is wildly popular within the English-speaking elite in India (versions have also been published in Hindi, Urdu, and Punjabi) and has achieved a cultural significance within India that possibly surpasses that of Roger Peterson's field guides in the United States, first published in 1934. Ali was part of an uncoordinated global phenomenon of ornithologists attempting to introduce the interested public to the subtleties of differentiating between species of birds through well-written and concise guidebooks with clear and simple illustrations.

Indian Birds was well written and conceived. For most it served as a useful introduction to the common birds of the subcontinent. Ali noted in his autobiography that U.S. and European servicemen in India helped sales immeasurably during World War II. And Ali, like Peterson in the United States, introduced many Indians to the hobby of bird watching. J. C. Daniel, honorary secretary of the BNHS, recalled that "forty years ago, if you told people you were going into the field to observe birds, they would laugh at you. But now there is a good consciousness in the country, at least among the educated people, about what you are doing. Institutions like this one have been important in educating people, and particularly, the old man himself, Sálim Ali. His name was always there, pushing for progress with birds and national wildlife matters. And birds are easy to watch, and they are everywhere."[44] Change a name here and there, and Daniel could be referring to any number of American or European field guide authors. Ali's book had a nationalist appeal, however, which guides in noncolonized nations lacked, and that gave his book a special significance to many middle- and upper-class Indians.

Before the book's publication at the height of the Indian nationalist movement, other scientific (not popular) guides to Indian birds had been published, but all had been written by British zoologists. As Ali describes, "Among the distinguished early admirers of the book have been Pandit Jawaharlal Nehru [later to be India's first prime minister], an ardent nature lover, who got a copy autographed by me while [he was] lodged in Dehra Dun [sic] jail (ca. 1942) to send as a birthday present to his daughter Indira [Gandhi, later a prime minister of India], then herself in Naini jail." Gandhi later wrote about receiving the book: "Like most Indians I took

birds for granted until my father sent me Shri Sálim Ali's delightful book
...and opened my eyes to an entirely new world." Nehru's fellow-prisoner
Maulana Abul Kalam Azad, then the resident of the Indian National Congress, reported having borrowed and used the book while in jail in
Dehradun.[45]

Ali's book was likely seized upon by people like Nehru, Azad, and the
young Indira Gandhi as part of a reclaiming of India by Indians. It is well
documented how intertwined the process of colonialism is with the acquisition of (often, scientific) knowledge about the colony.[46] The British
had been quite active in this regard, mapping India, studying Indian languages, religions, and customs, and cataloguing India's plants and animals—and then writing books to describe their findings and place
(British) order upon the world. Ali's book was the first such English-language
book to be written by an Indian.[47] Ali's book was really one of a pair; Stanley H. Prater, the Anglo-Indian BNHS curator and old friend of Ali's,
wrote the corresponding *Book of Indian Animals,* published in 1948 by the
BNHS. Prater had informally approached Ali about writing a popular bird
book as early as 1930, and Ali was formally asked to prepare the manuscript by the BNHS in 1935, at the same time that Prater was beginning
work on his own manuscript. Although Prater's book never achieved the
popularity of *Indian Birds,* it still has not been superseded as a general
guide to India's mammals. Its relative lack of popularity is at least in part
because mammals are much more difficult to observe than birds, but also,
I suspect, because Prater was not Indian and not as connected to the leaders of the Indian independence movement.[48]

Although Ali usually avoided political commentary within his autobiography, his life was very much intertwined with the political occur-
rences of the twenties and thirties. He mentioned in passing that his family
entertained Gandhi at one point, a brother of his conducted Congress's
investigation into the Jallianwalabagh massacre in 1919, and he was on
good terms with Nehru and Indira Gandhi throughout his life. After independence Ali's book continued to receive attention from the highest
levels of the Indian government, and when the tenth edition of the book
was released in 1977 Indira Gandhi was in attendance at the release party.
Ali was reclaiming India's very heritage through his work, a treasure that
had been plundered and claimed by the British during colonialism. Of
course, Ali had done his share of plundering in Rangoon, but by the forties he was a committed conservationist, and more to the point, he was an
Indian ecologist writing about India's ecology.

The Rise of U.S. Influence in the BNHS

One of the more surprising twists in the tale of the development of Indian ecology was the shift from primarily British to primarily American influence. Britain has a strong tradition of field ecology and British ecologists have been active throughout the world during the twentieth century. They have been leaders in the international environmental movement, including organizations like the WWF and IUCN. More specifically, the BNHS was for all intents and purposes a British undertaking in its early years, and British scientists had access, connections, and experience throughout India during the colonial period. The British Museum's collection of Indian bird skins is still larger than that of the BNHS. The occasional German interloper aside (as when Stresemann convinced Ali to bring birds to Germany), British scientists aggressively cornered the market on Indian ecological expertise. Ali states this quite directly, even claiming that the period of Indian ornithology before World War II should be named the Ticehurst-Whistler era, in honor of two noted British ornithologists who worked in India.[49]

Following World War II however, British influence within the ecological sciences in India dropped off dramatically. This can be partially attributed to the end of formal British colonialism in India in 1947; after this date many British administrators, scientists, soldiers, and workers in India returned home. Many British scientists decided to work in other regions of the world still under British control, most famously Africa.[50] Further, wealthy U.S. institutions willing to fund ecological research around the globe were emerging in the late forties, when Britain was rebuilding from World War II and still trying to hold onto a shrinking global empire with a strained national treasury.[51] In 1947, along with independence, India became a creditor to Britain, who had borrowed heavily from India's treasury to support the war effort during World War II. There was also an important personal dynamic at play: Sálim Ali was without question the key postindependence figure in Indian ecology, and his relationships with British scientific institutions (like the British Museum, which had refused his request to study there in 1929) and ecologists had at times been acrimonious. U.S. institutions and individuals carried no such colonial baggage for Ali.

Ali's difficulties working with British ecologists were typical for Indian scientists. When British ecologists came to work in India in the thirties and forties, they came as scientists, but also as privileged citizens visiting the empire in the midst of a liberation struggle. One of them, Col. Richard Meinertzhagen, while planning a 1937 trip to Afghanistan with Ali, rejected Ali's choice of K. N. Kaul to accompany them as a botanist: "He [Kaul] is a young man, nice mannered and intelligent, but I am a little doubtful whether I can stomach two seditionists for three months all day and every day. Sálim is a rank seditionist and communist, so is Kaul (a brother of Jawaharlal Nehru's wife) and it would probably end in disaster." At another point in his diary, Meinertzhagen wrote, "Like all Indians [Ali] is incredibly incompetent at anything he does; if there is a wrong way of doing things, he will do it, and he is quite incapable of thinking ahead." Still later: "He is prepared to turn the British out of India tomorrow and govern the country himself. I have repeatedly told him that the British Government have no intention of handing over millions of uneducated Indians to the mercy of such men as Sálim: that no Englishman would tolerate men being governed by rats."[52] In the twenties and thirties, Ali depended on British scientists for advice and funding for his research and thus he had to maintain a working relationship with them. However, it appears that he did not suffer their colonial arrogance in silence and that his work on Indian birds was carried out in a fairly tense political climate.

Not all British scientists were as disagreeable as Meinertzhagen. But in the forties the older generation of British naturalists, which had done so much in the first decades of the twentieth century, was retiring or dying, while the younger generation was leaving newly independent India to return home or to other portions of the crumbling British Empire. The older British ornithologist with whom Ali had the closest relationship, Hugh Whistler, had died during World War II of cancer.[53] Whistler had been a close friend throughout Ali's career, freely providing advice and support to Ali throughout the preceding twenty years. Ticehurst, the British ornithologist who had been so upset about the birds Ali took to Germany, had also died during the war. Meinertzhagen had retired by 1947. Ali had close relationships with many younger British scientists at the BNHS, but almost to a man those scientists returned to Britain in 1947 and 1948 and conducted little research in India after that. Ali's good friend S. H. Prater was the most important member of this group. Despite his being born in India, Prater was convinced that there was no longer any place for an Englishman in India. He gave up the curatorship of the BNHS to return to Britain in 1948, in spite of Ali's sincere entreaties to stay in India as the curator (while Ali would be the honorary secretary) and help run the BNHS. Had he done so, I would be writing a very different history of that institution. As it is, the end of 1948 found Ali solely in charge of the BNHS as both curator and honorary secretary, with his British mentor dead, abandoned by his British colleagues, and (I think we can safely assume) fairly nonplussed with the prospects of further collaboration with any of the old India hands from Britain.

Enter S. Dillon Ripley

Into this context, the tides of World War II had swept a young American to the beaches of Bombay and into Ali's life. S. Dillon Ripley had visited India as a young teenager in 1926–27 on a family tour. In 1943 he returned as a member of the Office of Strategic Services (OSS), the World War II precursor to the CIA. Ripley had finished his Ph.D. in biology at Harvard in 1943 while simultaneously working at the U.S. Museum of Natural History (part of the Smithsonian Institution). He was determined to help in the war effort but had been turned down in an attempt to join the navy. As soon as he had defended his dissertation, he promptly enlisted with the OSS as a civilian expert. Ripley was assigned to the secret intelligence branch and posted to Ceylon (now Sri Lanka). Ripley, who had participated in a scientific expedition to New Guinea in the thirties, in addition to his earlier trip to India, was considered something of an expert on South and Southeast Asia. By March 3, 1944, Ripley was named chief of secret intelligence for the OSS in Ceylon. Ceylon was the intelligence headquarters for Southern Asia—chosen over an Indian location because of the greater stability of the Ceylon government. As part of this work he oversaw operations and participated himself in brief missions to Thailand and Burma as well as many trips to New Delhi.[54]

Although he wanted to help with the war effort, the war was by no means his only concern. As he remembered in a later interview, "I went down *determined* that I would utilize my spare time properly, and make a collection of birds for the Smithsonian." In his free time in Ceylon, Ripley would travel through the countryside in a chauffeured military jeep, with a man on loan from the Colombo Museum of Natural History who would travel with him and serve as his "bird skinner." During the war, Ripley continued to publish scientific articles, including one that proposed a new race of nightjars (a nocturnal bird) he had collected, identified, and classified in between his war duties. Ripley was not alone in his obsession, though his academic training in biology gave him a certain amount of renown in the small circle of wartime birders. Ripley described one bizarre trip to Burma in which a British lieutenant general (who had heard of the American OSS

Figure 6. S. Dillon Ripley at his home in Litchfield, Connecticut.
Photograph by Harry Naltchayan, April 1969, housed at the Smithsonian Institution Archives

ornithologist) asked for help on a bird manuscript he was preparing on the birds of Burma.[55]

On one of Ripley's first trips to New Delhi in 1943, he arranged to travel through Bombay, again combining his OSS work with his ornithological passion. He hoped to meet the famous Sálim Ali, author of the weaverbird studies (which he had read) and of *The Book of Indian Birds* (which Ripley had purchased). This first attempt at a meeting did not succeed as Ali was out of town on a field expedition. After finding out that the young American scientist had been to Bombay, Ali wrote Ripley from the field: "I am so sorry to have missed you in Bombay, but I hope you saw Prater and were able to glance over our collections. I of course know of you being Assistant Curator of Birds at the U.S. Natural History Museum, but had no idea you were on a fighting job and out here. I am always most happy to make contacts with foreign workers, and hope it will be possible for us to meet here before you leave my country, whenever that may be."[56]

The two of them would indeed meet. As Ripley remembers their first meeting: "I first met Dr. Sálim Ali in Bombay—a diminutive figure, slim almost to the point of emaciation, clean-shaven at first, but now with a trim, pointed beard. The meeting was an electric one, for—above all— Sálim possesses ineffable charm and also a penetrating concentration on the life of birds." He also said of Ali, "I was attracted to him when I met him— he's very cheery, full of humor, and we became good friends."[57] Ali had read some of Ripley's papers (which he had been impressed with),[58] and was aware of Ripley's position in the United States, and they began a friendship that would last until Ali's death. Ripley considered Ali "my best friend in Asia," and on Ali's seventy-fifth birthday Ripley led a birthday celebration including ornithologists from four continents (they eventually published a volume of essays honoring Ali).[59] Ali wrote of Ripley that he was "a lifelong friend and valued colleague, fellow-ornithologist, and explorer." When I read their correspondence of forty-four years, it was impossible to not be touched by the depth of their friendship. After their first meeting, Ripley made it a point to travel through Bombay whenever possible for the duration of the war, and they "discussed the possibilities of joint ornithological fieldwork and specimen collecting in India, and built air castles about the expeditions we would make together once the war was over."[60] Their plans became reality in 1946, when they made their first joint field trip to the northeastern state of Assam—they would continue taking field trips together until 1982, when Ali was eighty-six (and Ripley was sixty-nine). Ripley was still trying to persuade Ali to make one more field expedition with him right up to 1987, when Ali died of cancer.

When Ripley met Ali, he recognized, as British scientists could not or did not, the great benefits of working with such a talented and knowledgeable Indian expert. At a time when Britain was still desperately trying to hold on to her empire, Ripley did not carry that burden in his interactions with Ali. Instead, he could sympathize with Ali. On his first journey to India, as a teenager in the twenties, he had been shocked by the Indian civil servants returning to their colonial station; "They were snobbish in the extreme and were horrified at the thought that some Americans had gotten aboard the boat in Egypt." Time and again during their stay in India, Ripley and his family had been treated by the British colonialists as not quite up to par. On his return to South Asia during the war as an OSS officer, Ripley found relations to still be strained: "We few Americans were considered to be suspect by the British authorities who were really concerned with the law and order and the traditional administration of India itself—suspect in the sense that they perceived America's entry into the war and activity in the east as a threat to the colonial empires of Britain, France, Holland, and they viewed us as interlopers in some way."[61]

Ripley was not alone in this perception. Many OSS officers serving in Asia noted that their European allies were worried about U.S. actions in the area. A history of the OSS notes, "American diplomacy quietly frowned on British imperialism, and some OSS officers informally opposed British moves they viewed as efforts to expand the empire."[62] In fact, much of Ripley's war work involved making sure that Thailand emerged from World War II as an independent nation rather than as another British colony or protectorate. The British were not particularly worried about Ripley himself; when Ripley saw the British spy report on him from the war, it said, "Ripley is not important. He is simply a gilded bird man."[63] Had they known the content of Ripley's reports to his superiors in Washington advising them on how to act in Southeast Asia to support nationalist movements, they might have been more concerned.

More than a shared distaste for colonialism linked Ali and Ripley. Unlike any of Ali's colleagues, Ripley had been exposed to the new ideas of the German bird behavior ecologists and trained in that tradition. As an undergraduate at Yale, Ripley majored in history and pursued ornithology as a serious hobby. While at Yale, Ripley first encountered the German behavioral ecology ideas when Erwin Stresemann himself came to New Haven for a one-year term. Ripley later recalled, "I wanted to meet the man who was then thought to be the most distinguished living ornithologist." Yale offered Stresemann a position, but he refused it and returned to Germany. As Ripley commented, "Heaven knows what would have happened if Stresemann had come, because he was such a seminal figure[.] . . . [I]f he had come to America he might have had even greater success and certainly a more comfortable life." Stresemann was not a Nazi sympathizer, but he never did leave Germany. After World War II, Ripley sent him food and clothing, and Stresemann, in a letter of thanks, commented on the troubled Germany in which he lived, but then wrote that, after all, "Ornithologists live a happy island life."[64]

Obviously, not all German academics shared Stresemann's contentment with an "island life," including one of his students, Ernst Mayr. American universities and labs absorbed an immense influx of German scientific talent in the thirties. The most well known of these were the nuclear physicists and mathematicians, but ecologists had fled Hitler as well, including the brilliant Mayr, who came to the American Museum of Natural History in New York City before eventually moving to Harvard several years later. It was at the American Museum that Ripley first met and worked with Mayr, whom he would eventually call his "mentor in bird studies and systematic evolutionary studies."[65]

When Ripley graduated from Yale with a history degree, he had not been sure what career he wanted to pursue. He briefly considered law at Harvard, but decided that was not at all interesting. Ripley was without doubt part of the East Coast elite, but he didn't feel that he quite belonged to it. His parents had been divorced early in his childhood and he was estranged from his father. His mother had lost a good bit of her money during the stock market crash. His family had many connections and he was in no danger of going hungry, but he "didn't have any real prospects of . . . a cushy job." After much thought, he eventually decided, "I might just as well fall back on biology or zoology, convinced that I would never really be able to earn a decent living, but at least I'd be happy doing what I felt comfortable with."[66] Ripley decided to settle for the genteel (and relative) poverty of the Ivy League academic and promptly enrolled in Columbia University to do graduate work in biology. The parallels with Sálim Ali are striking—Ripley was another young man from a prosperous and well-connected family who chose to pursue birds instead of wealth.

Ripley disliked Columbia. Within a couple of months of starting there, he received an opportunity (through family friends) to travel with a wealthy young couple that were sailing to New Guinea. They wanted a naturalist to travel with them. Apparently it was not much of a contest between New York City and the tropical islands of the Pacific: "I cast care to the wind, threw up my course at Columbia, went down to the American Museum of Natural History, told my new friend, Ernst Mayr, a young German student of Stresemann who was teaching and curating [there], that I had an invitation to go to New Guinea. He himself, as a graduate student in Berlin, had been to New Guinea, and he virtually pummeled me on the back and said 'Go to it—Marvelous!' So I spent from November until January studying with him, sort of going over lists of the fauna and the birds of New Guinea."[67] This was a turning point in Ripley's biological career for at least three reasons. Most important, it gave him a reason and opportunity to work closely with Mayr, both immediately before and after the trip. Under Mayr's tutelage Ripley became well versed in the most sophisticated bird studies in the world—the same theories and ideas that Ali had encountered seven years earlier in Germany. The New Guinea trip introduced Ripley to the study of Asian birds, the region and fauna that would become his life's work. And finally, the trip convinced him to leave Columbia. When he returned he would enroll instead at Harvard and finish his Ph.D. there.

Mayr had not yet moved to Harvard when Ripley was a student there. With both his undergraduate and graduate degrees, Ripley preceded the true ecology synthesis led by his two alma maters, Harvard and Yale. (This synthesis would begin in the mid-forties and combined the botany traditions of Clements and the Midwestern U.S. plant ecologists, the behavioral emphasis of the German bird biologists, the focus on ecosystems of Hutchinson at Yale,[68] and Mayr's neo-Darwinian ecology/evolution synthesis.)[69] Ripley even noted that when he took an ecology course at Harvard in the 1940–41 school year, "I liked it almost least of any of the courses I took during my graduate years, finding it insufficient, poorly taught, and essentially missing many of the main points."[70] This comment indicates not only the premature state of the science at Harvard in 1940 but also that Ripley had an idea of what it should have been—he knew what "main points" his class was missing. This was thanks to his work with Mayr. Although Mayr was not officially at Harvard, he remained in close contact with Ripley. In one letter during the middle of Ripley's Ph.D. work, Mayr wrote, "It is fortunate that you are up in Cambridge and not in New York, because I have brain-waves all the time concerning subjects that you might like to work on."[71]

Although Ripley missed a lot of Mayr's brain waves, he still caught enough to insure that his graduate training at Harvard was done in the context of the best German ecology. Ali and Ripley knew of each other before they had even met because they moved in the same small global circle of bird ecologists and read and wrote articles for the same journals. In a letter to Ripley during the war Mayr called Ali "the most gifted and competent of the Indian ornithologists."[72] Upon meeting they recognized in each other not only common friends and interests, but also a dedication to a similar vision of ecological work (what Thomas Kuhn would call a common paradigm).[73]

Ripley's status as an American immediately after the war would matter in one other way—as Europe focused on rebuilding itself, U.S. institutions were comparatively wealthy and the funding of international scientific expeditions was comparatively frequent and easy. Even if a postwar British scientist were eager to work in India, securing funding for what was now an ex-colony (as compared to many places in Africa, which were still under colonial rule) would prove very difficult. Similarly, newly independent India had many demands on its scarce financial resources, and bird studies were surely not near the top of the list. Ali knew this and moved to secure U.S. funding for his ornithological work in India. As Ali wrote to Ripley still during the war, "After the war and when you are back in the U.S. I hope we may be able to work joint expeditions of this kind on a larger scale. It is obvious that funds are much more easily available in America for this sort of scientific work, (and will be for some time, I expect, until we come into our own) [than] in India." Again, in a letter a year later as he and Ripley were preparing a trip to Assam, Ali wrote, "I am glad to hear you say that you would be coming out to India again. We have wonderful places for birding and collecting. Our great handicap here is funds. You provide the funds and we can do whatever you like together."[74] There is enough evidence from other sources to know that Ali did not suffer fools gladly, and I am confident that it was not money alone that made him eager to work with Ripley. When Ali offered to "do whatever you like together," he knew that the type of ornithology that interested Ripley would match his own interests. But the wealth of the postwar United States certainly did not hurt their collaboration in this early stage of their friendship.

When Ripley met Ali in India, he did not carry with him the assorted baggage of British scientists from this period. Instead, he carried training influenced by the German ecologists, an anticolonial politics, and access to the resources of postwar America. This combination, added to their personal compatibility, helped the two forge a partnership that assumed even more importance as they continued to rise to global prominence after the war.

Following the war Ripley returned to Yale as a faculty member, and by 1964 he had become the director of the Smithsonian Institution (in Washington, D.C.), a position he held until 1984. At the Smithsonian, Ripley oversaw hundreds of scientists working across the world, and he helped transform the Smithsonian into one of the most ambitious and successful ecological research and training institutions of the second half of the twentieth century.[75] Halfway across the world, Ali became the leader of the BNHS in 1947, when the British leaders of the society returned to England. Ali was uninterested in official titles and administrative work and he soon found others to become curator of the collection and honorary secretary. Ali eventually succeeded in surrounding himself with a capable and fiercely loyal group of young Indian naturalists and ecologists who saw to the day-to-day running of the society while Ali continued his fieldwork and wrote his books.[76] There was never any doubt where the real power of the BNHS lay, though, especially given Ali's lifelong friendship with Jawaharlal Nehru and Indira Gandhi, who were the prime ministers of India during much of the rest of Ali's career. By the time of India's independence, the BNHS was inextricably linked in the Indian public's mind (and in the reality of organizational structure) with Sálim Ali and with the study of birds. When Ali met Ripley, links were forged not just between two outstanding scientists but between the two most important natural history research organizations of their respective countries—the Smithsonian, a world leader in tropical ecology, and the BNHS, the most sophisticated indigenous ecological organization in the developing world.

The Ali-Ripley Collaboration

The most tangible result of the Ali–Ripley partnership was their coauthored ten-volume masterpiece the *Handbook to the Birds of India and Pakistan* (1968–74). Ali and Ripley were renowned ornithologists at the peak of their careers and this guide, thirty years in the writing, was the culmination of their shared professional ambition. It was a state-of-the-art summation of the taxonomy and ecology of every known bird species on the Indian subcontinent, and it looks to maintain that position well into the future.

Ali had proposed that they work on a collaborative project as soon as the war was over. Ripley, still a young scholar, was not yet set on how his career would proceed. Ali was determined that Ripley should work on Indian birds. Ripley received similar advice from Mayr: "So far as I know, there is nobody left in England except Kinnear who knows or cares about the taxonomy of Indian birds."[77] Mayr went on to suggest that Ripley should work with Ali. Ali and Ripley made their first field expedition together in 1946 in northeastern India. On that trip, Ripley had brought nylon mist nets to use in their collecting (these nets were at that time rare, and this was their first use in India, but they have since become standard ornithological equipment).[78] It was the first in a long line of technical innovations supplied to Indian ecologists by U.S. scientists, who had greater access to equipment and to funding. (Later innovations included jeeps, radio collars, tranquilizing darts, and GPS systems).

When Ali and Ripley used the nets they discovered that the published range and distribution estimates for many species of birds were wrong.[79] Ali pointed out to Ripley that an entirely new technical ornithological reference was needed to supplant the old Fauna of British India series, one that included both new taxonomic and speciation data, new range and distribution data, and discussions of the ecology of each species.[80] Further, the guide should not only be "for the use of professional museum ornithologists but also for the amateur bird watcher, and therefore well-illustrated and with popularly written life histories of the birds . . . and an authentic source of complete information about all that had so far been recorded on some of the species, and what aspects were not sufficiently known and needed further study" (180–81). The two of them could work together and do something significant, Ali argued. Ripley agreed.

In planning the book, they decided that Ripley would focus on the taxonomy and systematics and Ali would focus on the ecology. This division reflected their particular strengths—Ripley's work on Indian birds had primarily been taxonomic, while Ali had countless hours of observation in the field—but it also testified to Ali's reputation as an ecologist, not just an identifier of birds. This mutually agreed upon division of labor reversed the typical north/south ecological divide of the twentieth century in which ecologists in the developing world time and again have been encouraged by the north to simply catalogue biological diversity, or to act as field guides for visiting northern scientists, and to leave the teasing out of meaning, theory, and behavior to the ecologists from the developed world.[81]

Ripley would have enjoyed working on some of the ecology himself but recognized that as a young assistant professor at Yale it would be difficult to do so without much more time in the field. Later in his life Ripley looked back on this decision with bemusement: "I *did* divert in the sense that I went back to doing more collecting. I got more interested along the way in doing rather large monographic treatments of the birds of India and things of this sort, which were less in the field of behavior or speciation—perhaps more in old fashioned systematics. . . . Perhaps I relapsed into the more classic taxonomic mode."[82] Ripley felt that doing this taxonomic work was a trade-off. It allowed him to make collecting trips to exotic locales (much theoretical ecology takes place in labs, and he did not have the time for extended behavior studies unless they were conducted in zoos), and he knew that he was contributing to an overall work that was much more than just taxonomy. In fairness, during the twentieth century the classification of birds in a taxonomic schema was supposed to reflect evolutionary connections, and what made two similar birds different species, or merely races of the same species, was highly contested and reflected much more scientific thought than simply cataloguing species might suggest. And Ripley was consciously doing taxonomical work in the interest of the new ecology, not as an end in itself.

It is good that Ripley could be interested in taxonomy. Ali hated it. In just one of the dozens of similar rants in his correspondence with Ripley, Ali complained, "My head reels at all these nomenclatural metaphysics! I feel strongly like retiring from ornithology, if *this* is the stuff, and spending the rest of my days in the peace of the wilderness with birds, and away from the dust and frenzy of taxonomical warfare. . . . The more I see of these subspecific tangles and inanities, the more I can understand the people who silently raise their eyebrows and put a finger to their temples when they contemplate the modern ornithologist in action."[83] Not surprisingly for one who did not enjoy taxonomy, Ali was bad at it. Mayr cautioned Ripley, "I don't think he ever understood the necessity for collecting series. Maybe you can convince him of that."[84] As Ali's collaborator, Ripley simply took care of the collecting himself on their numerous joint field trips.

In the initial years of their collaboration, Ripley began to work on placing all the Indian bird species into the most modern taxonomic order, and Ali continued to collect behavioral observations. First, Ripley would need to complete his taxonomic work, a checklist of all the Indian birds, which would eventually be published separately as *Synopsis of the Birds of India and Pakistan*. Only after this work was published (so that other ornithologists could critique and accept its new taxonomic designations) could Ali and Ripley then work on the handbook, in which Ali's ecological information would be placed in Ripley's taxonomic categories.

Ali alternately praised and prodded Ripley from afar, trying to assure that the younger man stayed on the straight and narrow. In 1947, Ali wrote Ripley, "I am glad you resisted the invitation to Kenya. I do not want you to divert from the field of Indian birds." Later in the same letter, he looked forward to their next chance to work together on a field expedition in India. Ripley's energy rose and fell—the summer of 1949 must have been a particularly low period, with notes from Ali stating, "You must cease loafing," and then a month later, "I see that you are in the midst of your annual summer rut. I hope you will get over it soon because there is a good deal of more useful work to be done. I wish you would put off the writing of popular books until you finish the check-list." Ali actively worked to raise Ripley's profile as an expert on Indian avifauna: "Have you any matter for publication in our Journal? I would welcome it as I am anxious that your contributions on Indian birds should become more and more known in India." There were other scientists working on Indian birds right after the war, including a young Indian who had briefly traveled to the United States to work with Mayr. With Mayr's implicit support, he was proposing his own version of a checklist for India's birds. Ali disliked this scientist's tendency to create multiple new subspecies with minute differences that Ali thought made fieldwork more difficult. Besides, Ali had decided that Ripley was to be his chosen collaborator, and he used his considerable influence to see that Ripley's version of the tax-onomy of Indian birds was the authoritative one—as he wrote Ripley in 1956, when Ripley was in the midst of a debate about the classification of a few species: "I am prepared to sympathize with you in all your worries and predicaments on this score, and hope that even under these circum-stances you will succeed in producing some order out of the chaos." Fi-nally, in 1961, Ripley's *Synopsis of the Birds of India and Pakistan* was published, its final completion spurred (as it often is) by the need for a significant publication for his academic promotion at Yale.[85]

From the end of the war to 1961, Ali and Ripley had gone on numer-ous collecting trips in India, and Ali, on a Fulbright fellowship, had enjoyed an extended stay at Yale with Ripley. Ali was busy throughout this period gathering ecological information on the numerous bird species of India. He had quickly realized that the task was "quite staggering, as I discover more and more, how exceedingly little we know about many of our birds. . . . No serious ornithologist has bothered to learn much about *living* birds."[86] After the publication of Ripley's *Synopsis,* Ali and Ripley could begin to move ahead and piece together their handbook. The writing it-self began in 1964, the year Ripley had moved to the Smithsonian, and proved much quicker than the collecting of field data and taxonomic work had been. The first volume of the *Handbook to the Birds of India and Pakistan* was published in 1968 and met with great approval.[87] Four days before the release of the tenth and final volume, Ali turned seventy-eight. Although he would live for another thirteen years, his writing was largely over, ex-cept for popular pieces focused on conservation and his autobiography, *The Fall of a Sparrow.*

The Ali and Ripley *Handbook* is an exhaustive source meeting the highest standards of their discipline. It is unequaled in Indian ecology for its com-prehensive coverage of a section of Indian zoology. This set of books was not only a scientific encyclopedia for ornithologists, but also a handbook for those committed to the appreciation and preservation of nature—en-vironmentalists. It also has great, if not as readily apparent, cultural signifi-cance. Mahesh Rangarajan comments, "We see here a broadening of horizons as these men cross boundaries of nation and color to work in a common endeavor of discovery." Mary Louise Pratt, in her work on impe-rialism and natural history writings, describes interactions between West-ern scientists and people in colonized nations as a "contact zone" that includes "coercion, radical inequality, intractable conflict, [and a] lack of

Figure 7. Sálim Ali and S. Dillon Ripley in Calcutta near the end of their long and fruitful collaboration. *Unknown photographer, 1979, photograph on file at the Smithsonian Institution Archives*

shared language." This description seems to fit the interactions between Ali and the British Colonel Meinertzhagen to a tee. The two men were constantly at odds, and Ali was in a subordinate position, dependent on Meinertzhagen for his job and access to the field project. And though the two men shared a grammatical language, English, they spoke a very different scientific language. Meinertzhagen was a British taxonomist, cataloguing the empire, while Ali had by that time been trained as an ecologist and was interested not just in taxonomy but also in interactions.[88]

In contrast, Ripley and Ali did not meet in a contact zone but on the common ground of a shared understanding of ecology, a shared interest in birds, and a mutual respect for each other's strengths as scientists.[89] The two men shared a common vision of what ecology was, a vision largely resulting from the work of German ornithologists in the early years of the century. That they shared this vision was the result of the effect of global politics on the development and spread of a new science; Ali ended up in Germany with Stresemann when the British would not have him due to the politics of colonialism and the effects of the global economic depression, and Stresemann's student Ernst Mayr emigrated to the United States, where he subsequently trained Ripley. But for these events, Ripley might have met Ali (itself a result of a random U.S. military assignment sending

Ripley to South Asia), but never collaborated with him because they would have had different paradigms as to what constituted worthwhile and interesting ecological work. As it is, they understood and liked each other's work, and the initial connections were made that would lead to the growing influence of U.S. ecologists in India in the following decades. When the British left India (and abandoned the BNHS) in 1947, Ripley, followed by other U.S. ecologists, was in position to help fill the breach with funding, collaboration, and technical expertise, further cementing the development and spread of Ali's preferred science of ecology, versus taxonomic zoology, as the ascendant version of Indian natural history.

Chapter Three

Looking for the Jungle

U.S. Ecologists in India

> I say that we have just begun to explore the world. It is still possible for any individual to go on the equivalent of a nineteenth-century expedition to a valley in Ecuador or a peak in New Guinea and find new species of butterflies, to go to islands that have been virtually unexplored for whole sections of the living world. . . . I don't think the dark forest will ever be fully explored.
>
> —*E. O. Wilson, "Ecology and the Human Imagination"*

S. DILLON RIPLEY, WHO CAME TO INDIA to work with Sálim Ali from the 1940s through the 1980s, was the first of a stream of U.S. ecologists coming to India to do fieldwork, replacing the British scientists who had once been ubiquitous in both the forests of India and the pages of the *JBNHS. Replace* is perhaps too strong a word, for it was clearly different for U.S. scientists to work in the independent nation of India than for British colonialists to survey their empire. However, the warmth and personal rapport of the Ali–Ripley collaboration was not typical for most U.S. work done in India in the sixties through the eighties. Like British scientists before them working in India, the U.S. ecologists had funding, equipment, and training that were not readily available to their Indian counterparts. This disparity meant that Indian and U.S. ecologists often did not meet as equals.

Ali was the exception, not the rule, among early Indian ecologists. He was supported by a relatively wealthy family through the first forty years of his life, during which, for all intents, he earned no money but devoted himself single-mindedly to bird ecology.[1] He had exposure to cutting-edge ecological theory in 1929–30 in Germany. He also had the good fortune to be born in Bombay with an uncle who was a member of the BNHS and could write him an introduction to the staff. And finally, his friendship with the Nehru-[Indira] Gandhi family, the center of nationalist political life, did nothing to hurt his influence in and access to decision-making and policy-setting power. This is not to discount Ali's accomplishments—privilege alone has never determined innovation—but the very specific type of privilege Ali enjoyed was a precondition for his success in the early years of the Indian republic. I am not aware of any Indian ecologists similarly trained, funded, or placed before the late sixties and a scant few after that.[2] In his last letter to Ripley, written in the spring of 1987, just a few weeks before Ali's death, he brought up the question of who might be able to revise the *Handbook* that had been the culmination of his and Ripley's scientific careers:

> [It should keep its] emphasis mainly on ecology and bionomics, by a person who has had sound field experience of birds in the Indian subcontinent. . . . He would need to keep up to date with international literature on species on the Indian list and update and make appropriate changes where necessary. I cannot, at the moment, think of any experienced Indian ornithologist who would have the time or dedication to undertake the responsibility. . . . This is a matter that we must attend to as quickly as possible since I feel that I am definitely drifting towards the abyss and . . . I would not like to think of all the good being interred with my bones![3]

There was no one to whom Ali could pass the torch as the next birdman of India. For all of his success as a scientist and an environmental advocate, even by the eighties Ali had failed to establish disciples among his many supporters and followers in the BNHS with the same degree of broad ornithological expertise that he had achieved himself.[4]

After Ripley, when U.S. ecologists came to do fieldwork in India with scientists other than Ali and on subjects other than birds, they came not as colleagues but as "experts." Unlike the British scientists who had worked in India previously, though, most of these U.S. ecologists came as trained ecologists (not taxonomists), and they came with the encouragement and even active recruitment of the Indian staff of the BNHS. The BNHS saw these U.S. scientists as possible means of furthering their conservation goals and increasing their store of scientific information, and U.S. scientists saw in India the opportunity to explore new (to them) biota and to work in exotic and diverse locales, on species ranging from lions and tigers to rhinoceroses and antelope.

Why Did the BNHS Want U.S. Ecologists to Come to India?

Ali and the leaders of the BNHS were very aware of the state of Indian ecological science in the first decades of Indian independence. India had, and still has, an amazingly rich biological diversity—it is one of the world's "megadiversity" nations.[5] For most of India's species of plants and animals, though, there was little or no ecological data in the 1940s. The state-of-the-art summary of India's animals at the time of India's independence was Prater's *Book of Indian Animals*. The entry on the jackal, one of the most common, widely dispersed, and easily observed of India's wild mammals, is typical of the knowledge of that time: "Little is known about the Jackal's family life—it is so secretive in habits. How long the male remains with its mate or family, what part, if any, it plays in caring for the young, their upbringing, growth, and dispersal, all remain to be discovered."[6] About more elusive animals even less was known, if that is possible.

This lack of knowledge was given urgency by what some Indian observers perceived as an environmental collapse in the years immediately following Indian independence.[7] With independence, private ownership of guns had proliferated and more Indians were hunting for sport and food than ever before. Many Indian citizens rejected "shooting regulations as a form of colonial restraint and . . . shot down wildlife everywhere, including sanctuaries and private estates."[8] Added to this was India's rush toward industrial development and agricultural self-sufficiency, which resulted in many Indian forests and undeveloped lands being converted to pastures and fields.[9] For a few conservationists, the euphoria of independence was followed by a mounting fear that the environment of the subcontinent was being irreparably harmed. In this context, the fates of three charismatic large mammals were particularly alarming environmentalists: the Asian lion had become restricted to the Gir forest in Kathiawar District in the state of Gujarat, the Asian one-horned rhinoceros was restricted to a small population in northeast India and Nepal, and the Indian cheetah was suspected to be extinct by the early fifties, last spotted in 1948, and officially declared extinct by the sixties.[10] Ironically, no one was concerned about the tiger yet.

The leaders of the BNHS were concerned about India's growing environmental crisis, but they did not feel they had enough information to definitively advocate conservation policies. Working from the assumption that ecological knowledge was a precondition to any conservation initiative, in the late fifties and the sixties the society explicitly set about trying to encourage detailed studies of Indian ecology. Ali himself pushed the BNHS to begin a massive bird-banding project in India, a project that required large infusions of outside money (almost completely supplied from the United States), but for which Ali already possessed the necessary expertise.[11] The BNHS was poorly equipped to carry out ecological studies involving animals other than birds, however. For those studies they needed expertise as well as funds. Here, the BNHS chose to actively support ecologists from outside India wishing to work on nonavian as well as avian Indian ecology. This support took the form of providing affiliations for foreign graduate students, helping to secure research visas, selecting local guides or field assistants, and offering the *Journal of the BNHS* as a place for publication of their research.[12] For ecologists who came into India independently of the BNHS, staff from the society still would contact these people, solicit their advice, visit them in the field, and encourage them to publish in its journal.

The BNHS thus took a very pragmatic stance: foreign researchers offered the best means for increasing what was known about Indian ecology and for exposing young Indian ecologists to the most current techniques and theories of the science. The *JBNHS* did not try to compete with international journals like *Ecology* or *Nature,* instead accepting a more local status. This enabled visiting scientists to publish results in the *JBNHS* in addition to the well-known international journals. In turn, Indian scientists who could not afford subscriptions to more expensive Western journals still would keep up with cutting-edge ecological research (as long as it occurred on the subcontinent). Unlike many other developing nations in which ecologists came, collected results, and returned home, never to be heard from again, ecologists working in India actively contributed to local dialogues about Indian ecology.[13] Knowingly or not, these U.S. scientists were also contributing to the development of a cadre of Indian ecologists who often took their initial inspiration from the pages of the *JBNHS.*

There were two additional benefits that Ali and his colleagues saw in encouraging U.S. ecologists in their work in India. Foreign scientists were often given greater attention by the Indian press and Indian government officials than Indian ecologists. In a 1965 ceremony dedicating a new BNHS office building (Hornbill House), the union minister for education, Mr.

Chegla "remarked that his own eyes were opened to the wealth of wild life in our country when it was pointed out to him by foreigners who were always puzzled by our apathy to it." The BNHS skillfully used foreign visitors to increase the profile of environmentalism in India. Second, the BNHS was able to fund its increased environmental outreach in India through foreign research collaborations. The minutes of the annual general meeting for 1970 indicate that "the reason why the society was in a satisfactory position financially was . . . that a sum of Rs. 24,982.52 accrued as administrative fees for handling various projects from grants received from the Smithsonian Institution, Washington, Yale University, and other sources." This influx of foreign money was a major source of funds for the BNHS through the fifties, sixties, and early seventies. Immediately after independence, Sálim Ali had succeeded in getting a temporary grant from Prime Minister Nehru to support the BNHS when many British members had stopped their financial support. The BNHS had to turn to other sources to make up their budget (as the government grant was only temporary), and research collaborations with wealthy U.S. institutions were a major source of funds.[14]

What the BNHS Got, or Two U.S. Ecologists in India

Many U.S. ecologists came to India to do work in the sixties and seventies and have subsequently become influential within Indian conservation biology circles. Steve Berwick first came to India from Yale as one of a small group of graduate students doing research in the Gir forest on the ecology of the last home of the Asiatic lion. He helped organize a conference held in India in the eighties on wildlife ecology techniques that many practicing Indian ecologists cite as a turning point in their careers (both with regard to acquiring new techniques and new contacts), and he coedited an influential book attempting to adapt wildlife biology techniques to the Indian context.[15] Michael Fox, who came to India in the seventies to study the dhole (a wild dog), introduced A. J. T. Johnsingh to ecology. Johnsingh went on to become the first Indian ecologist to do a Ph.D. on an Indian mammal (the dhole); he is now head of the Wildlife Institute of India's biological sciences division (where he was the advisor for Williams's elephant research). Fox returned to India to work in a wild animal veterinary clinic.[16] A surprising number of U.S. scientists came to India and made a real difference through their mentoring, their publications and research, and their environmental advocacy. Here I focus briefly on two ecologists who came to India in the sixties before Berwick and Fox, George Schaller and Juan Spillett. The stories of these two men, one a world-famous ecologist, the other a lowly graduate student, illustrate how U.S. ecologists were received in India and suggest what was distinctly different about their work than that which had occurred before.

George Schaller

More than once, George Schaller has been called the greatest naturalist of the twentieth century, most visibly by his famous admirer Michael Crichton.[17] Schaller came to India on September 2, 1963, to conduct research on what was then called the Bengal tiger, now simply the Indian tiger, and its associated prey species. Although only thirty years old, Schaller was already well known for his work in Africa on the mountain gorilla.[18] The book he wrote after his year of research in Kanha National Park in the state of Madhya Pradesh in central India, *The Deer and the Tiger,* was not as popular internationally as some of his other books, but it was arguably his most influential scientific work.[19] Schaller forever changed Indian wildlife biology.

As I traveled across India in 1998 speaking with ecologists, Schaller's name came up again and again in response to my standard question about significant scientists or studies that had influenced their life and work. Senior and junior ecologists alike mentioned Schaller with deep respect. One flatly told me, "George Schaller is a god." A. J. T. Johnsingh stated, "I largely take ideas and inspiration from George Schaller. . . . He was one person who called himself a naturalist, but he also started the first scientific study of wildlife in India." J. C. Daniel, after mentioning Sálim Ali's work in ornithology, said, "As far as studies that have been done in India in the last forty years, Schaller's work, *The Deer and the Tiger,* was the baseline. He was the one who really started field studies on a sustained basis. I would say that he was the first important person in conservation biology." Ullas Karanath, an ecologist who studies tigers in southern India for the World Conservation Society (formerly the New York Zoological Society), an organization for which Schaller is the chief biologist and Karanath is the only Indian staff scientist, told me, "I was always interested in tigers. In the mid-sixties, when I read George Schaller's article [in *Life* in 1967], I said, 'This is exactly what I want to do!' . . . I said 'This is a step ahead from merely describing and saying the tiger is beautiful or saying so many tigers were shot—going beyond that, looking at prey numbers.'" Raman Sukumar, an elephant expert at the Indian Institute of Sciences, paid homage by having Schaller write the preface to his book *Elephant Days and Nights.*[20]

During a midday rainstorm in the Kalakad rain forest, a graduate student at the Wildlife Institute told me about how she made a fourteen hour (one-way) bus trip for the chance to hear Schaller speak. Afterward she told him about her project radio-tracking small mammals. "I hope you're not using satellite telemetry," he commented. "No, radio tracking." "Good," he replied. "Good fieldwork must be in the field, not at a computer terminal." Her story finished, she turned away from me and looked back out the open door at the streams of rainwater running past the little house that was her field base, across the mud that was her footpath, and down the mountains on a rainforest floor that seldom saw the sun. Schaller is more than just a model for many Indian biologists. He is a source of inspiration and an example of how an ecologist can, through solid fieldwork and good writing, change the world in which he or she lives.

Schaller had received funding for work in India on the parasites of wild ungulates from the Johns Hopkins School of Hygiene and Public Health, working out of the newly established Johns Hopkins Center for Medical Research and Training in Calcutta. This study was funded by the U.S. National Institute of Health, and lasted from his arrival in 1963 until his departure in May 1965. His subjects were the various wild deer species of central India, with special attention to both visible and microscopic parasites. His research involved tranquilizer darting to take blood samples and collecting macroparasites, and he collected deer feces for analysis by other scientists. Johns Hopkins University, with the establishment of its Calcutta field station in 1961, had decided to have an active presence in India. Although the Calcutta center was medically oriented, it also had an ecological unit headed by U.S. ecologist Charles Southwick.[21] Public health researchers were deeply interested in the ecology of wild animals because of the possibility they could serve as vectors for human disease. Animal-to-human transmission (e.g., bubonic plague, African sleeping sickness) made compelling arguments for investigating whether other links might be found.

Many U.S. ecologists in India and elsewhere received funding from the National Institute of Health or other public health programs and they have done exactly what Schaller did: fulfill the grant requirements but also conduct a field-based study on something else. Schaller was given funding to research Indian ungulates, specifically the chital: "At first I intended to study primarily the chital deer, but it soon became apparent that the most fruitful approach, the one that would yield the most information not only for the program but also from the standpoint of conservation and management, would be to collect a broad spectrum of facts on a number of different species" (8).

I doubt whether Schaller ever truly intended to study just the chital deer. He wanted a research project with teeth, literally. Schaller changed his research topic to include the Bengal tiger and its prey species (the prey was what Johns Hopkins and the NIH were interested in). Schaller realized that no scientific studies of the tiger had been done. Even more significant, Schaller approached his study not as an investigation of one species (the tiger or, as his funding agencies might have suggested, the chital) but as a study of the interrelationships between the assorted large mammals of Kanha National Park—a study of the park's ecology.

Schaller wanted to work in a study site without livestock and with a low human population, part of the typical U.S. rationale in selecting a study site, as I will discuss below. After over a month of searching for a research site, Schaller chose Kanha, "a relatively undisturbed forest area" (8). Other areas were eliminated from consideration because "the animals in the unprotected forests were usually so . . . sparse," and "the sanctuaries of Rajasthan supported a vast livestock population." Still other parks were not satisfactory because "for six months of the year the grasses are ten to fifteen feet high in many places, making it difficult to see the animals much less to observe them at length" (12).

The majority of Schaller's work at Kanha was observational, usually from the roof of his Land Rover (14). He recorded over two hundred hours of sustained observation on each of the ungulate species he studied and, with the assistance of cattle staked out as bait, managed to spend several nights observing tigers (although they weren't hunting or eating naturally) (15–16). Schaller made vegetational transects to analyze the plant species composition and abundance so that he could better estimate what the ungulates were eating. He also analyzed kill sites and feces of the tigers and ungulates. Mostly, though, he watched through his binoculars. As with Sálim Ali, Schaller had been trained in the ecological tradition and looked for interactions and behavioral patterns in his chosen study species. It was the way he watched, not the watching itself, that differentiated Schaller from early naturalists in India like Jim Corbett, who spent countless days and nights observing tigers (first in order to kill them, later to photograph them). One reviewer described Schaller's book as "important because it records the observations of a well-trained biologist."[22]

While in Kanha, Schaller was constantly in touch with the staff of the BNHS. Scientists visited him in the field, and he made brief visits to other parks to assist Indian foresters and conservationists in population counts. One of these visits was recorded by J. C. Daniel in the *JBNHS;* the article is effusive in its praise of the "American Ecologist." Daniel relied on Schaller to show him and his team the most effective means of surveying a park to estimate the population of wild buffalo.[23] This was an example of the BNHS scientists taking advantage (with Schaller's best wishes) of a visiting ecologist who is persuaded to share with them the latest field techniques and to assist them in collecting valuable field data on population numbers. Population numbers were important, of course, as indices of the need for greater efforts for conservation. The numbers were given greater weight because they were produced in collaboration with a "world famous" ecologist who was using state-of-the-art scientific methods—not just guesswork.

The scientists of the BNHS met *The Deer and the Tiger* with unqualified praise and admiration.[24] The review in the *JBNHS* was glowing. The reviewer, E. P. Gee, was a noted environmentalist and a member of the IUCN as well as a close friend of Ali and loyal member of the BNHS. He was quite direct in placing the book within the context of other such studies: "For the first time on this subcontinent, a dedicated scientist has remained almost continually for 14 months in what is possibly the finest remaining natural habitat for wild life found in Asia. . . . The book contains a wealth of factual data which almost bewilders the reader in the revelation of the amount of time and hard work involved."[25]

Gee had visited with Schaller while Schaller was in the field and was quite impressed that he was willing to stay in the park throughout the monsoon, when Kanha is cut off from the outside world. Most ecologists who had previously worked in India had been unwilling to endure such hardships. Sitting in a blind with Schaller while a herd of chital (spotted deer) moved across a meadow before them, Gee watched Schaller record group size, habits, feeding preferences, and interactions. He was amazed at how much information Schaller could extract from what appeared to be a relatively uneventful hour. This, Gee concluded, was what U.S. ecology had to offer India—the ability to derive valuable information about India's wildlife in a systematic and scientific fashion on the basis of seemingly innocuous animal behavior that was opaque to the untrained eye. Gee concluded in his review, that Schaller's book "should be read and studied by every forest officer, both before and after taking charge of a division."[26]

Whether or not every forest officer has read the book, it is almost a certainty that every Indian ecologist has read at least portions. Inspired by

Schaller's work, Gee began to conduct tiger censuses throughout India. On the basis of the shockingly low numbers he found, he began advocating increased tiger conservation. Gee's efforts, in conjunction with those of other ecologists and environmentalists (both Indian and Western), led to a IUCN conference in New Delhi in 1969 that both raised global awareness of the plight of the Indian tiger and helped establish an Indian chapter of the WWF.[27] This conference, and the WWF, then proved crucial in the formation three years later of Project Tiger.

Juan Spillett

Schaller's story is, like Ali's, atypical in that it is a story of a truly exceptional scientist who worked far beyond the normal boundaries of his peers. The vast majority of scientists do work of an entirely different type, work that Thomas Kuhn calls "normal science," the necessary working through of the details sketched out by larger paradigms or theories. Kuhn argues that all scientists spend most of their careers conducting normal science, and only occasionally do a few break the bounds of the established scientific paradigms.[28] For instance, Ali's two papers in the 1930s on the Baya weaverbird and on bird pollination of flowers were pathbreaking, but his later and more famous work with Ripley for the ten-volume *Handbook* was normal science—the working through of details of bird ranges and behaviors, not the creation of new concepts or paradigms. There was much more of this normal science occurring in India in the sixties than there was revolutionary work such as *The Deer and the Tiger,* with its novel focus on the interactions between predators and prey and its implementation of new field techniques in the study of Indian wildlife. Ph.D. student Juan Spillett's work in India falls into the category of solid, normal science. A brief look at Spillett's work with the BNHS helps to reveal how less-famous U.S. ecologists were received by the BNHS and Indian conservationists.

Spillett was an American who went to India to do his fieldwork in the mid-sixties when he was a Ph.D. student at Johns Hopkins, having already earned a master's degree in wildlife resource management at Utah State University. The vast majority of U.S. ecologists who worked in India over the last forty years have been graduate students.[29] This is not surprising; graduate students are the workhorses of scientific research. This is especially true for ecological field studies, which often require that the researcher stay at the field site for a protracted period. The differences between field ecology and "bench" biology (that which occurs in a lab) are fairly obvious. University professors, having done their field adventure as graduate students, typically find it difficult to leave their other responsibilities and go to the forest for a year or two; they might visit the field site as advisors at the beginning and end of the research.

Spillett was funded by the Johns Hopkins School of Hygiene and Public Health and was affiliated with the Centre for Medical Research and Training in Calcutta, as was Schaller. Spillett had even visited Schaller and assisted him for a portion of his work at Kanha.[30] Originally, Spillett was to have carried out his Ph.D. research on the black buck, an antelope tigers prey on. At the last minute, however, Johns Hopkins decided that ungulate studies were of limited use (Schaller's work had not turned up much in the way of public health connections). Thus, Spillett found himself conducting research on the lesser bandicoot rat in Calcutta grain warehouses. This was urban work that clearly had public health value (this rat carries fleas, which carry the plague bacillus) and was also of economic concern (even today this species eats a vast amount of stored grain).[31] It was not what Spillett wanted to do, however. For a wildlife ecologist looking forward to studying antelope in the forests of India, rats in Calcutta are a sorry substitute.

The BNHS came to Spillett's rescue. E. P. Gee had been working since the First World Conference on National Parks (Seattle, July 1962) to hire a trained ecologist to conduct a count of the highly endangered Asian rhinoceros in Kazaringa Wildlife Sanctuary, in the state of Assam. After persistent lobbying, Gee received funding from the WWF in 1966. Gee knew that Spillett was working in Calcutta and he quickly arranged for Spillett to use the WWF funding to conduct a survey not just of Kazaringa and not just of rhinos, but of all the mammals in all the wildlife sanctuaries of northern India. For six months Spillett traveled "over 13,500 miles by every conceivable form of transport from the aeroplane to the bicycle and the boat and covered 300 miles on foot and spent over 21 days on elephant back."[32] On these surveys, Spillett was primarily concerned with censusing some of India's more important mammal species, and surveying the general ecological condition of the parks. His schedule required him to spend only a few days at each park before traveling to the next. After completing his report, Spillett returned to his rats, only to leave them for another few months in 1967 to conduct a quicker and more informal survey of the sanctuaries of southern India. Surely the whirlwind BNHS tours were welcome breaks.

Obtaining accurate information on the population sizes of rare animals in a sanctuary was quite important. Most simply, the number (or estimated number) of a species tells naturalists if they need protection. In the absence of reliable data, it is impossible to bring any pressure to bear on a government to take action. In a 1964 review of E. P. Gee's recently published *Wildlife of India,* Ali spoke of the need for "an experienced and competent expert of the International Union for the Conservation of Nature to be invited to India to do some pioneering in wild life census work here, and also to train local workers in the adaptation of approved scientific techniques to local conditions."[33] That was exactly what Spillett was being asked to do.

The BNHS strongly felt the need for an expert to help them with population figures, so that they could push their conservation agenda, and Spillett fit their bill perfectly. He had been well trained as an ecologist. Furthermore, he was in India, he was available, and his words would carry the cachet of coming from an "international" expert. The December 1966 issue of the *JBNHS* was devoted exclusively to Juan Spillett's reports on his wildlife surveys and to an article on the chital by two U.S. scientists working in Hawaii. In the lead editorial, the editors wrote, "One of the problems of the conservationist in India, is the lack of authentic knowledge of the status of species and habitats in our country." This volume of the *JBNHS* was meant to be an example of the sort of work done by "expert" ecologists (all from the United States), and the results that could be gleaned from such work. It was quite unusual for the *JBNHS* to publish an article that was not the result of fieldwork in India, but they justified the Hawaiian research by claiming that the species was the same as the chital of India and that their "main object in publishing this very comprehensive report is to draw attention to the type of work that has to be done in this country if we are to be in a position to assess our wildlife resources properly."[34]

Although Spillett spent only a few days in each of the parks he visited and in the parks of southern India he only had time to talk with the forest officers, he was able to overlay his observations in the field with knowledge he had acquired elsewhere. For instance, when dealing with the dangers of livestock grazing in the sanctuaries, he wrote, "with the abuse of overgrazing, the most palatable or desirable plants are the first to disappear . . . until eventually all that remain are plants that the animals would not normally eat." Spillett could not observe this occurring in India, but, based on his earlier work at Utah State, he was aware of this dynamic. Of course, any ecologist would be aware of the dangers of drawing assumptions from one part of the world (e.g., the North American

Rocky Mountains) and applying them to another (the grasslands of India's central Deccan plateau). But given the time constraints and the desires of his BNHS hosts, Spillett did the best he could. And he did not shy away from making strong policy recommendations to India's political elite that exactly corresponded with the BNHS position: "India has been richly endowed with precious natural resources. Many of these, however, already have been destroyed or lost due to ignorance, tradition, apathy or political expediency. . . . Unless the leaders of India are soon able to implement definitive measures and initiate sound conservation practices, little more than want and poverty and the eventual weakening of this great nation can be expected."[35]

As an outside expert, Spillett met the tactical needs of the BNHS. His report lent credibility to their claims of imminent environmental collapse, and it provided preliminary population estimates of highly endangered mammals. Thanks to the BNHS, Spillett had an opportunity to make policy recommendations from a highly visible pulpit within India, a role he certainly could never have filled in the United States as a graduate student. Although Spillett's data was highly unreliable (short field visits lead to bad data), his work was implicitly understood by the BNHS to be more of a starting place than an endpoint. In any case, Spillett was accompanied in all his travels by assistants from the BNHS, who attempted to learn as much from him as they could in the field so that they could replicate his surveys with more detail and more time.

This was not a new move on the part of the BNHS. As early as 1949 records show Ali using Ripley to generate "expert reports" for BNHS advocacy goals. In a letter to Ripley, Ali asked him to write a report recommending that sanctuaries be established in Assam to protect rhinos. Ripley was in Connecticut at the time and could not do fieldwork, so Ali told him what to say: "perhaps stress the point even further that all outside activities in the sanctuaries should be promptly and totally stopped, i.e. grazing, fishing, etc." Ripley apparently responded to Ali's satisfaction, and Ali moved quickly to use this latest piece of expert international testimony, advocacy without evidence. Only three weeks later Ali wrote Ripley, "I have suggested to Dr. Hora . . . to sponsor a resolution on behalf of the Bombay Natural History Society morally charging the Assam government to take all necessary measures as recommended by the Bombay Natural History Society's experts [Ripley] to save the Great Indian One-horned Rhinoceros from extinction." Spillett was in good company.[36]

Why Did U.S. Ecologists Go to India?

What the BNHS hoped to gain by encouraging and supporting U.S. scientists working in India is clear: funding, up-to-date theory, new technologies, cosmopolitan contacts, and international advocates. But why did so many Americans want to come? While India is a mega-diversity nation, so is the United States. Further, a good case could be made that there are more "wild" areas left in the United States than there are in India.[37] And it is certainly less expensive and more convenient to do fieldwork in America. But come to India the young ecologists certainly did. For example, in its annual report for 1971, the BNHS reported,

> The Society maintains close contact with the International Union for the Conservation of Nature and the World Wildlife Fund as well as the Smithsonian Institution. There are many researchers abroad who would like to be sponsored by the Society to do research work in India. The following list will give an example of the type of proposals we receive: The Ecology of the Gir Leopard, Chicago University; Comparative Studies in Evolutionary Ecology, Smithsonian Tropical Research Institute; Niche Ecology of the Garden Lizard (*Calotes versicolor*) in the Gir Forest, University of Kansas; Ecology of Indian Crocodiles, New York Zoological Society.[38]

Two things stand out from this listing. First, all the proposed projects are from institutions in the United States. Second, the U.S. institutions are diverse in type and location, including a public and a private university, a private organization (the New York Zoological Society), and a national research institution.

Nor did U.S. ecologists head just for India. Beginning in the early twentieth century, but accelerating rapidly after World War II, U.S. ecologists (initially, field biologists) spread throughout the world. Before the end of the war, U.S. ecologists predominately worked in U.S. colonies (Panama Canal Zone, the Philippines, Puerto Rico) and other tropical nations in Latin America. Following the war, they were encouraged to travel throughout the world, particularly in the many nations newly rid of European and U.S. colonialism or newly dependent on U.S. economic or military aid. One example at the institutional level was the Pacific Science Board, formed by the National Research Council after the war. Under the leadership of Harold Coolidge, of Harvard's Museum of Comparative Zoology, the board's express purpose was "reopening the vast reaches of the

Pacific to scientific research."[39] The militaristic language—"reopening"—points to the top-down (not mutual!) bent of U.S. government attempts to encourage the study of the ecology of the world. The Pacific was also the recipient of some of the fruits of U.S. science that were less than peaceful, as with the military's tests of nuclear weapons.

Most U.S. ecologists needed no such institutional encouragement, though funding and institutional connections certainly were not spurned. In the post-1945 United States, field ecology was wrapped in an aura of discovery and adventure. And it was a growing discipline. Between 1945 and 1966 the number of members of the Ecological Society of America quadrupled.[40] The science of ecology was just becoming codified in institutional structures as university departments and hiring lines in the fifties and sixties, and correspondingly the number of students studying ecology grew rapidly.[41] The Organization for Tropical Studies was founded in 1963 as a consortium of universities and scientific research institutions to encourage the training of young field biologists.[42] With the attendant boom in the environmentalism movement, the growth of ecology programs in the U.S. accelerated throughout the sixties and seventies. For young U.S. ecologists, ecological fieldwork combined the glamour of scientific research, ascendant as never before in the post-Hiroshima world, with the culturally loaded opportunity for solitude, wilderness exploration, and discovery on an (ecologically) uncharted frontier.

Wilderness, the Frontier, and Ecology

There is a well-known tradition of naturalist exploration in the West, dating at least to the Spanish, Portuguese, English, French, and Dutch investigations of the Indies and Americas in the sixteenth through the nineteenth centuries. Darwin, with his trip around the world in the 1830s, is the most recognizable of the explorer-naturalists, but he is only one among hundreds of European men who felt the thrill of discovery in the attempt to catalogue and "know" the newly colonized world.[43] There was a briefer tradition of European naturalists in the North American colonies than in many Old World areas (like the British in India), but there was the compensating lure of the still uncharted and unclaimed American West for U.S. naturalists in the early years of that young nation. Whether exploring the Louisiana Purchase, searching for dinosaur bones, or simply striking out into the "wild" like John Muir, aspiring U.S. naturalist-explorers found ample room for discovery on the North American continent.

Historian William Cronon suggests that American cultural ideas about the frontier merged with romantic ideas about "sublime wilderness" in the mid-nineteenth century to produce the belief that "wild country became a place not just of religious redemption but of national renewal, the quintessential location for experiencing what it meant to be an American."[44] Scholars within American studies (as well as other sites in the academy) have long investigated the frontier and wilderness as mythic symbols within American culture.[45] Cronon combined these two strands of analysis to argue that it was the wildness of the frontier, particularly its vision of large stretches of nature with no human touch or habitation, that gave (and continues to give) the frontier such power in the American imaginary. By entering the frontier, an explorer or settler was also entering the wilderness, a place of solitude and sacredness in which a new self could emerge. Further, "in the wilderness, a man could be a real man, the rugged individual he was meant to be before civilization sapped his energy and threatened his masculinity."[46]

Late-nineteenth-century U.S. naturalists who struck out for the "wilds" were not just walking in the wild, then, they were entering a crucible from which they would emerge fully realized individuals: wiser, purer, and manlier. This is the very myth of Theodore Roosevelt—the city boy who became a man in the wild frontier of the Dakotas in the late 1800s and came to be seen as the manliest of American politicians, a strong and forceful leader.[47] Not incidentally, he is often remembered as one of the first U.S. presidents to actively promote wilderness preservation. Of course, it was only possible for East Coast Americans and recent settlers to imagine that an unpeopled wilderness existed across North America by either removing the American Indians from the landscape or ignoring their presence as "savages," part of the wild itself.[48]

U.S. ecology emerged from its traditions in natural history at the beginning of the twentieth century, mirroring a similar movement in other parts of the world. Ecologists in the United States, however, were steeped in these American cultural dialogues about the importance of wilderness and frontiers, ideas not as prevalent elsewhere. Early U.S. ecologist Frederic Clements is an example of how this mattered. Clements was the creator of the climax theory of vegetation. Though most ecologists now disagree with crucial elements of this theory, it is strongly entrenched in popular environmentalism to this day. This theory holds that all ecosystems have a dominant community of plants that constitutes its natural and undisturbed climax state. Any disturbance to that ecosystem (logging, fire, grazing, etc.)

will disrupt this climax, but over time if the disturbance ceases the climax will reemerge through a set succession of states recapitulating the set in the previous cycle. For instance, logging wipes out part of a climax forest, which then is left as grassland. Over time, shrubby plants, which will eventually give way to larger trees, will colonize this grassland. Thus, the climax forest will eventually return. The climax community was assumed to be the most suited vegetative community for any ecosystem.

This theory justifies a belief that in the long term the wilderness comes to rest naturally in a climax community (contemporary ecologists prefer to think of the successional stages of a forest not having any set end point, instead being in a continuous process of change). Wilderness is scientifically codified by this theory as stable and unchanging, once a climax state has been reached. The climax community is by definition peopleless, and any human activity (other than traveling through its midst and quietly observing, we might assume) is classified as disturbance—disturbance (temporarily) destroys the climax community. Further, the theory suggests that if people are removed from a terrain, the ecosystem will eventually return to a wilderness state. This notion authorized the belief that the American West would return to a natural state after the removal of the American Indians and that wilderness sites in the East would eventually return to their prehuman climax community if left undisturbed by people. It is difficult to prove that cultural beliefs solely are responsible for scientific theories and policy prescriptions, but it is at least possible to surmise that the wild popularity of Clements's ideas in the face of conflicting theories and evidence reflects the power of U.S. cultural beliefs and the confluence of his theory with those beliefs.[49] In the decades following Clements, U.S. ecologists have consistently oriented their science to the study and preservation of unpeopled wilderness—a term defined by cultural, not scientific, standards.

U.S. cultural beliefs about wilderness mattered to ecologists in another way. As settlements spread across the American landscape, there were fewer and fewer places to which an explorer could go to truly be immersed in what felt like the frontier—indeed, the 1890 census had officially declared the U.S. frontier to be "closed." When the U.S. Navy extended its reach throughout the Pacific and Caribbean in the 1898 Spanish-American War, some American field biologists began to extend their research programs to these newly acquired tropical colonies in search of untouched wilderness and new discoveries—what William Cronon calls Americans' "deep fascination for remote ecosystems."[50] Remote ecosystems became U.S. ecologists' wild frontiers. U.S. scientists often idealized these remote ecosystems as "wild," with inhabiting peoples either romanticized as wild parts of the ecosystem itself (as with aboriginal peoples), or dismissed as nuisances to be avoided and ignored at all costs. Similarly, these areas were explored as uncharted frontiers—obviously, they were frontiers only to the visiting ecologist, not to the people who already lived there.[51] The dismissal of local peoples and local knowledge also allowed ecologists to make easier discoveries of "new species," and certainly, they were new species to Western science. Of course, some species, particularly insects, were perhaps undifferentiated by local peoples; they might have lumped a couple of different species of ants together as one type of animal in their forest. And one Indian scientist, trained at the University of Florida, told me, "I had a tribal guide who tried to tell me that the racket-tailed drongo [a long-tailed black bird] was the male and the tree-pie [a long-tailed orange, white, and black bird] the female of the same species."[52] Neither species of bird was unknown to local people, but they were known differently.

As American imperialism expanded overseas at the end of the nineteenth century, ecologists marched alongside. By 1927 the Smithsonian Institution alone had sponsored over fifteen hundred scientific expeditions "covering the globe." Listing the institution's activities, its assistant secretary offered, "It procures foreign diplomatic and learned recognition and assistance to expeditions going abroad. Credentials addressed to 'Friends of the Smithsonian Institution' insure an open door even in semi-barbarous countries." No doubt many of the wild frontiers explored by U.S. field biologists were in these "semi-barbarous" countries—clearly placing U.S. ecologists and institutions like the Smithsonian in the context of contemporaneous European imperial activities throughout the globe. At times these field biologists overstepped their accepted boundaries, as when Reginald C. Harris became involved in an insurrection of local Indians against the government of Panama—at such times the Smithsonian did its best to disavow itself of any responsibility, for they wanted to be on good terms with even semibarbarous governments.[53]

One of the most prominent examples of the expansion of U.S. military might early in the twentieth century led to the creation of the world's most famous tropical research station, known the world over as BCI. When the United States completed the Panama Canal in 1914, the diked-back waters created islands out of former hilltops. One such new island, Barro Colorado Island (BCI), was set aside for scientific purposes by the

Canal Zone's governor in 1923. The eighty-nine-hundred-hectare island was formally turned over to the Smithsonian in 1946, though scientists from the institution had been running the research station there since its inception. It was renamed the Smithsonian Tropical Research Institute in 1966. Later a sizable area (3,900 hectares) of the mainland near the island was also acquired by the Smithsonian and converted into part of the Barro Colorado Nature Monument.[54]

The station is completely self-contained and operates, to all intents and purposes, as a U.S. field station. At one point or another it has supported practically every tropical ecologist from the United States—most of whom were trained there as graduate students participating in summer learning programs generously funded by the Smithsonian's Tropical Research program.[55] In 1979 alone "nearly 1,200 biologists from twelve countries came to the station to conduct their researches[.] . . . [M]ore researchers are now being turned away than can be accommodated."[56] I have even known of researchers, with grants in hand, who were unable to go to BCI during particularly busy seasons. For many young scientists, this station provided their template for how one conducts research overseas— a template that included no foreign governments, no natives, a protective "moat," and a largely U.S. staff. The research conducted at BCI for decades has been voluminous and important, and BCI's presence has facilitated a phenomenal growth of attention and interest in tropical ecology within the United States. Although the Panama Canal has been returned to the Panamanians, BCI will belong to the Smithsonian for the foreseeable future, a little piece of the United States in Central America.[57]

Following World War II, U.S. ecologists, almost exclusively men, ventured forth in even greater numbers into a world made much more accessible by America's unprecedented wealth and power. BCI is clearly the most extreme example of a tropical "world made accessible," but it was by far not the only place easily available to young U.S. ecologists carrying U.S. dollars. When hundreds of young men entered U.S. ecology graduate programs after World War II hoping to conduct specific types of fieldwork in the tropics, it is fair to wonder why. I think that the confluence of American beliefs about unpeopled wilderness, unexplored frontiers (preferably remote), and the linking of masculinity with risk and discovery in the wild, offer clues. For many of these young men, like E. O. Wilson, ecology offered an opportunity to explore unknown wildernesses and through "a test of my will severe enough to be satisfying . . . know myself better as a result . . . and go home."[58]

It is difficult to spend much time socially in a room full of field biologists without the conversation working its way to anecdotes of near-death experiences in the field. A U.S. woman who received a master's degree in wildlife biology in the late seventies referred several times to the machismo that permeated the university where she had studied. According to this informant, grizzly bears ("they call them g-bears") were considered the most admirable species to study, while her own work on woodpeckers was rated quite low on the macho index. For her peers, field ecology was, at least rhetorically if not in reality, as much about testing yourself in situations of great danger as it was about acquiring information about all elements of nature, great and small. This is not an artifact of this one university; it is a common theme in my discussions with ecologists throughout the United States. This is also a common element in popular understandings of wildlife biology, as witnessed by the popular TV show *Crocodile Hunter* on the U.S.-based Animal Planet channel. On this show, a roguish and ecologically knowledgeable young Australian man consistently places himself in positions of great danger vis-à-vis very large carnivorous animals, ostensibly to illustrate natural history lessons. Wildlife biologists, then, need only a dangerous study animal for adventure. For ecologists focusing on ecosystems or interactions between less glamorous subject animals, though, the need for a proving ground was provided by a foreign wilderness rather than just a dangerous study animal. Thus, if you study insects, as did Wilson, you find your adventure by international travel to remote locations.

Ultimately what a society considers manly is a product of culture and can be shown to change over time and across cultural boundaries. Athletics was commonly considered to be an arena for the demonstration of manliness, but the entry of millions of women into athletics has changed the gender-specific tone of athletic competition in America. The U.S. women's soccer team—highly successful, highly competitive, and highly physical—clearly is not perceived as demonstrating their manliness. Just because tropical field biology in remote ecosystems was an opportunity for young American men to prove themselves does not mean that there were not women with similar interests in adventure, travel, nature, and discovery. Until recently, though, women who might have been interested in field ecology did not enjoy as much access to training and funding, and thus until the last few years female tropical ecologists have been rare. There have always been women who were naturalists, but they have been a distinct minority, especially within professional ranks and they have been

Figure 8. S. Dillon and Mary Ripley on a field expedition in India, standing in front of the elephants and drivers who will take them through the jungle.
Unknown photographer, 1976, photograph on file at the Smithsonian Institution Archives

botanists, entomologists, or ornithologists, studying stationary or small or-ganisms of limited range and danger. They also have tended to work closer to home.[59] Many other women went to the field, but in the "his-torical shadows," as the wife of a renowned naturalist. Schaller and Spillett were accompanied by their wives, and Sálim Ali's wife used to accompany him, until she died early in their marriage. Mary Ripley accompanied Dil-lon on some of his field expeditions—Ali remembered her: "She struck me as singularly misfit for the rough-and-tumble conditions she would have to face. However, after sharing several strenuous expeditions with the Dillon Ripleys subsequently, I realized how sadly I was mistaken: Mary Ripley amid the elegance of her Washington drawing room is not the same as Mary Ripley in safari outfit in a dripping leech-infested camp."[60] Mary Ripley, Ali might have forgotten, had served in China during World War II, after all.

Even if a woman received training as a field biologist, securing a uni-versity position (a precondition for most practicing ecologists) in the sci-ences was nearly impossible for much of the twentieth century.[61] Within the U.S. academic community there was a definite bias against women. This was even more pronounced in remote field labs, which could feel like a "man's retreat." The one notable exception to the scarcity of women in field biology is the tradition of women primatologists who have worked in Africa and Asia, including Dianne Fossey, Jane Goodall, and Sarah Hrdy, among others.[62] These women, who began their work in the sixties, are heroes for many younger women who are currently striving to follow in their footsteps and literally enter the field.

Although change is occurring, it is slow. When I was at the Iowa Lakeside Lab in the summer of 1997, undergraduate and graduate students were taking field courses there. All the professors that summer were men, yet over 70 percent of the students were women (including all ten students in field ecology). It will take a while for those demographics to work their way into the academy. In the 1986 National Forum on BioDiversity, a hugely influential national conference sponsored by the National Academy of the Sciences and the Smithsonian (and where the word *biodiversity* was first used), there was one woman among sixty presenters. Mary Harris said that around 1987 she was the only woman among several hundred participants at a meeting of the southeastern branch of the Entomological Society of America. Nevertheless, I have listened to young women, in India as well as the United States, enthusiastically discuss large, dangerous study animals and their excitement when hiking through uncharted forests. As more women enter field ecology and press to participate in personal quests for wilderness, adventure, and self-discovery, this activity formerly glossed as "becoming a man" will acquire a different cultural meaning.

E. O. Wilson

E. O. Wilson is one scientist whose early career exemplifies the intersection of the cultural motivation of young American ecologists to travel to India and U.S. field ecology immediately after World War II. Wilson has never conducted field research in India, but he is well known and admired by ecologists there—one of whom is a former student of his. He entered graduate school in ecology in 1949 and has written a remarkably open autobiography, *Naturalist,* expressing his desire to combine his love of unpeopled wilderness, his desire for surveying exotic and new biota, and his quest for discovery at the scientific frontier. He feels he can, and did, do this through his career as a field ecologist: "evolutionary biology has become the last refuge of the explorer naturalist."[63]

Wilson sees nature as of necessity being peopleless. In his autobiography he writes of how as he grew up, "I turned with growing concentration to nature as a sanctuary and a realm of boundless adventure; the fewer people in it, the better. Wilderness became a dream of privacy, safety, control, and freedom. Its essence is captured for me by its Latin name, solitudo" (56). Of a field expedition, he says, "the solitude, as usual, felt right. Human beings mean comfort, but they also mean a loss of time for a field biologist, a break in concentration and . . . always a certain amount of personal risk" (172). In New Guinea, Wilson writes, "I was also hampered by the oppressively curious and helpful Hube people . . . who pressed in so closely to see what I was doing that I could barely work" (188). Wilson is extraordinarily courteous in personal interactions, and I suspect that he would be similarly polite to people who were disturbing his fieldwork. It is comical to imagine Wilson, aching to explore the wilds of New Guinea for new species of insects, surrounded by a horde of Hube villagers watching his every move, helping him turn over every rock. Clearly, fieldwork for Wilson best occurs in as deserted a place as possible, and ecology gave him license to make a career of heading away from people and into the wild. [64]

Wilson does not hide the fact that he is attracted to the new and unfamiliar. In one tropical forest, he recalls, "These creatures were a fully alien biota, and it is time to confess: I am a neophile, an inordinate lover of the new, of diversity for its own sake" (171). He always preferred the foreign animals at the zoo and as a child he played, "imagining myself to be in the jungles of S. America" (69). This theme of exoticism had a specifically tropical aspect that repeats throughout Wilson's autobiography.

Wilson craved this exotic, unpeopled nature so that he could explore it and claim a kind of primacy. Describing a different part of his collection trip to New Guinea, he writes, "I felt like a real explorer. I *was* an explorer, at least in the world of entomology" (183, emphasis in original). After he had climbed a tall mountain there, he wondered with a self-conscious racism, "Why did I come here anyway, except to be able to say I was the first white man to climb the central Sarawaget? That prideful goal was part of the truth, but I was looking for something more. I wanted the unique experience of being the first naturalist to walk on the alpine savannah of this part of the Sarawaget crest and collect animals there" (194).

Wilson was able to achieve his wilderness and his unexplored frontier by discounting non-Westerners, and he was able to be a discoverer by defining his exploration in terms of insects. Although Wilson is best known as the author of the controversial *Sociobiology* and as one of the ecologists responsible for the theory of island biogeography, he "was still possessed by an elemental self-image: hunter in the magical forest: searching not just for animals now but also for ideas to bring home as trophies. . . . The boy inside still made my career decisions: I just wanted to be the first to find something, anything, the more important the better" (210). Field ecology offered him many opportunities for firsts: "The Sarawaget, cold and daunting, had proved a test of my will severe enough to be satisfying. I had reached the edge of the world I wanted, and I know myself better as a result. By passing from the sea to its peaks I had finally encompassed the serious tropics of my dreams, and I could go home" (196).

The most striking statement of Wilson's ecological wanderlust and desire for wild, unpeopled frontiers is a prayer he wrote. He calls it his "archetypal dream":

> Take me, Lord, to an unexplored planet teeming with new life forms. Put me at the edge of virgin swampland dotted with hummocks of high ground, let me saunter at my own pace across it and up the nearest mountain ridge, in due course to cross over to the far slope in search of more distant swamps, grasslands, and ranges. Let me be the Carolus Linnaeus of this world, bearing no more than specimen boxes, botanical canisters, hand lens, notebooks, but allowed not years but centuries of time. And should I somehow tire of the land, let me embark on the sea in search of new islands and archipelagos. Let me go alone, at least for a while, and I will report to You and loved ones at intervals and I will publish reports on my discoveries for colleagues. For if it was You who gave me this spirit, then devise the appropriate reward for its virtuous use. (171)

Because unexplored planets, in this solar system anyway, do not teem with life and are rather awkward to reach, ecologists such as Wilson have settled upon the, to them, unknown lands of the tropics. Within these tropics, the fewer people that were found, the better. Where there were people, the ecologist-explorer should do his best to ignore them. It is noteworthy that while Wilson wants to explore alone, carrying his collection boxes, he still wants to engage in science—characterized here as the codification of knowledge shared with and critiqued by colleagues in some distant center. Still, though, this is a very solitary vision of science. He wishes for his primacy in discovery to be recognized, but that is the limit of his interactions with colleagues. Wilson's prayer differs from statements of imperial explorers in his lack of interest in claiming the land for the imperial center—he wishes to claim the land's creatures for Western, specifically Linnaean, science.

I do not believe that Wilson is unique among U.S. ecologists in his response to and feelings about unpeopled wilderness, new frontiers, and discovery. I think that he is unique in his ability to write so clearly about his desires, in his courage to be so honest, and in taking the time to do so in the midst of an astoundingly busy scientific career. The desire for an unpeopled nature was almost universal among the U.S. ecologists with whom I spoke.[65] This is evident even more tellingly in the journals of ecology: humans appear only as scientists, not as components of the ecosystem. You only have to look as far as the sheer number of northern ecologists who work in the tropics and actively advocate the preservation of tropical forests to realize that for many ecologists the forests are greener and the species diversity richer south of the U.S. border. And the thrill of discovery is present for every scientist, I believe. Ecologists differ only in that their discovery so often occurs outdoors while physically traveling and exploring, something they share with geologists but not with chemists, molecular biologists, or physicists. The "nature" that is relevant to this type of discovery strengthens the desire among U.S. ecologists for exotic field sites, where the chances of a new discovery increase. An untouched tropical rain forest is to a U.S. ecologist what a never-before-read archive of documents is to a historian.

We usually do not enjoy being told that our decisions are the result of cultural programming rather than our own self-interest or preferences. This is the cultural flip side of the common scientific argument—and the one made most strongly by Wilson himself, who sees even love of nature as in our genes—that we are all "genetic" dupes, fulfilling the evolutionary drives of our genes.[66] Totalizing arguments in either direction are not flattering, though humanities scholars often find culture an acceptable overlord, and many scientists seem somewhat comfortable with genetics in that role. With my analysis of the cultural motivations for U.S. field biologists, I do not mean to imply that tropical ecologists are cultural dupes. After all, I also crave walks through unpeopled tropical landscapes and, as an undergraduate biology major, pondered the possibility of naming an insect. However, it is useful to look for larger historical and cultural factors to explain repeated patterns of similar decisions made by unrelated individuals over a specific period. If a group of bright young American men chose to become field biologists and consistently acted, and explained their actions, in a very similar fashion, might there be some cultural reasons to explain this? Although the American fascination with remote wilderness and frontiers as realms of adventure, manliness, and self-actualization does not completely explain the actions of field ecologists in the mid-twentieth century, this U.S. cultural context, I would argue, provided ample reason for young ecologists to travel throughout the world, including to India.

In the decades following India's independence there was a coincidence of interests between young U.S. ecologists and the Indian ecologists at the BNHS. The former were in search of foreign lands and new discoveries, and had access to the money to pay for this search. The latter wanted access to the latest ecological theories, techniques, and equipment; needed money to support their own work; and hoped to tap the expertise of foreign visitors to increase the visibility of their environmental activities. These two groups initially came together through unique historical circumstances linking two important figures, Ali and Ripley. After their successful collaboration, the BNHS turned to the U.S. in the 1950s and 1960s for the funds, theories, and expertise it lacked. All of these things—as illustrated by the work of Schaller and Spillett—were forthcoming to the BNHS. Work in the jungles and forests of India, in turn, satisfied the desires of young U.S. ecologists looking for new biota and new experiences. And although the percentage of U.S. ecologists who worked in India is a small fraction of the total number of U.S. field ecologists from that period, they played a key role in Indian ecological sciences.

It is tempting to understand U.S. ecologists working in India as neo-imperialists. After all, the cultural background to U.S. ecology is strongly implicated in colonial practices and ways of seeing the world and obscuring other peoples. An imperial reading of U.S. ecologists working with the BNHS would be misleading, though. The Indian scientists of the BNHS *invited* the U.S. ecologists to India and supported them once they were there. As illustrated by Spillett's story, the scientists of the BNHS used some U.S. ecologists in a very calculated fashion to further their own scientific and political objectives. The BNHS scientists were able to use the funds they earned from collaborating with U.S. scientists to further their environmental education goals in India and to keep their society financially solvent. And, most intriguingly, the scientists of the BNHS could perceive their work with U.S. ecologists as a turn *away from* colonial scientific power relations, as exemplified by their relations with British scientists prior to independence in 1947. Choosing to work with the American ecologists rather than the British was liberatory for Sálim Ali and his colleagues at the BNHS. Because Ali had been trained in a mode of ecology unfamiliar to many of the British scientists in India, when he collaborated with Ripley he was conducting science that bore little relationship to what he had experienced as colonial science. The same U.S. ecological motivations and actions that could be experienced or read as imperialist in a Latin American nation subject to U.S. scientific (not to mention political and military) domination for more than a hundred years did not read the same way for Ali and his colleagues in the BNHS. And, after all, there were no Barro Colorado Islands in India. A more compelling argument could be made that the BNHS collaborated with U.S. ecologists in studies and policy recommendations that were imperial with regard to rural Indian villagers living in and around India's forests. The divide suggested by this would be one not based on nationality, but on the presence or absence of shared scientific, urban, and environmental values.

CHAPTER FOUR

Scientists or Spies?

Ecology and Cold War Suspicion

It is understood that U.S. officials, both in and out of this country, with the help of the American mass media made concerted efforts . . . to build up the image of Dr. [John] Seidensticker and expedite the long denied entry of foreign wild life "experts" to this country. It is believed that these "experts" want to use the strategic Bay of Bengal region for their own nefarious designs in the innocuous name of wildlife research. Some of them have a shady background also. Mr. Michael Huxley, science administrator of the Smithsonian Institute who once tried to come to this country for a similar "research," is known to have close CIA contacts. According to reliable sources, his mother is a CIA agent.

—The Patriot *(Indian newspaper), 1974*

NOT ALL OF INDIA IS TROPICAL. The small town of Bharatpur, for example, is roughly the same latitude as Palm Beach, Florida, yet in late December the temperature can remain in the forties in the day and hover near freezing at night. Although Bharatpur is in the famed desert state of Rajasthan, it is on the very eastern edge (about 150 km. south of Delhi), hundreds of kilometers from the desert, and drizzle or dense fog is common. As cold and miserable as Bharatpur can be in the winter, it is an avian paradise. Every fall, as frost spreads down from the North Pole to cover the forests and plains of Siberia and the states of the former Soviet Union, a colossal migration occurs. Storks, ducks, raptors, and songbirds take to the sky and fly south, past the Himalayas, past the Afghan hunters who are not above an occasional stork stew or roast duck, and past the winter wheat fields of northern India, until they reach a refuge of shallow open water in no danger of freezing and with plenty of fish. When they find such a place, the flocks land and make it their wintering grounds.

The most populous (in avian terms) of these wetlands in India is the group of marshes and wetlands covering twenty-eight square kilometers called the Keoladeo Ghana National Park, adjacent to Bharatpur.[1] By late December, over a quarter million[2] migratory birds have descended upon Bharatpur and joined its sizable year-round resident bird populations. Compared to Eastern Europe, Siberia, or the high Himalayas the birds find it balmy.

This avian paradise appears to be far from the realm of international politics and an unlikely place to begin a discussion of Cold War tensions or Indian fears of U.S. neo-imperialism. But in December 1967 Sálim Ali and his assistants were here capturing birds, banding their legs, and taking blood samples. The samples were screened for viruses and the data sent to the current funding agency for Ali's bird-banding study—the U.S. Army. A few years later, when the Indian press heard that Ali and the BNHS were being funded by the U.S. Army to conduct studies of migratory birds—birds that migrated to the Soviet Union and China, birds that could potentially carry pathogens (biological warfare agents) to these communist states—why then, Bharatpur was ground zero of the Cold War in India.

Based on Ali's enthusiasm for the migratory birds found at the wetland, Bharatpur had become the site of one of the most well known and important studies undertaken by the BNHS, a massive bird-banding project to map the migration patterns of India's birds. It was the most extensive and complex ornithological research attempted in India to that date. In the interest of sustaining this expensive project over three decades, Ali accepted funding from a variety of sources—the priority was to keep the project going. Unforeseen by Ali, the eventual results of his research were not just ornithological data, but also government inquiries into the BNHS, accusations of treachery against Ali himself, and a significant reduction in the freedom of U.S. scientists and organizations to work on and fund ecology projects in India. In the late sixties and early seventies, Bharatpur in December was not cold only because of the weather.

Before Bharatpur: The Politics of Indo-U.S. Ecology in the 1950s

Since the earliest days of the newly independent Indian republic, India and the United States have had at best a complex political relationship. India has been the world's largest democracy since its inception, and this has been a natural link between the two nations. At the same time, however, India purposefully steered an independent course during the fifties and re-

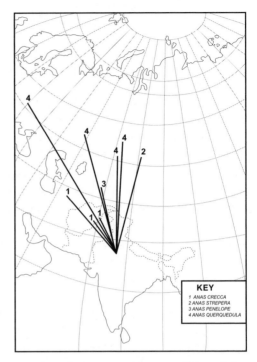

Map 2. Duck migration patterns.
Adapted from Lalitha Vijayan, *Keoladeo
National Park* (New Delhi: World Wildlife
Fund-India, 1994), 18, courtesy Orient
Longman Private Limited

KEY
1 ANAS CRECCA
2 ANAS STREPERA
3 ANAS PENELOPE
4 ANAS QUERQUEDULA

fused to slip into the dichotomized with-us-or-against-us politics of the Cold War United States. India's first prime minister, Jawaharlal Nehru, was one of the driving forces behind the Bandung Conference of 1955, which gave credibility to the idea of a nonaligned movement among the world's newly independent Asian and African states. Politically and culturally, leaders like Nehru were clearly oriented toward the democratic traditions of the West. At the same time, India shares a border with China and is very near the former Soviet Union. The new Indian government felt they had to develop independent political relationships with these powerful neighbors. Nehru's refusal to follow the U.S. line in global politics led to a certain degree of ambivalence in U.S. political views of India. India was often contrasted unfavorably with its smaller neighbor, Pakistan, which gladly allied itself with the United States in exchange for military aid. India was not treated as an enemy, though it clearly was not an ally either. Henry Kissinger, no great fan of India, believed that the U.S. government mistakenly treated India as a neutral arbiter of world affairs during the fifties, as the moral leader of the developing world.[3] Kissinger would have preferred a much more skeptical stance, and he helped to implement just such a policy during Nixon's presidency.

India was similarly ambivalent about the United States. While the Indian government wanted and needed U.S. economic aid, they did not want to become a U.S. dependent. Newly freed from British colonialism (only the latest in centuries of foreign rulers), the Indian leaders were determined that India should steer its own course in world politics and economics. The government of India restricted the entry of foreign companies into India, preferring to develop Indian industry internally. Nehru had visions of India being a regional power in Asia, and being a world leader in opposing imperialism.[4] Nehru saw Cold War alliances with the U.S. as beginning a new form of dependency; in a statement on Pakistan's alliance with the United States, Nehru said, "It would be unfortunate that Pakistan should gradually lose her independence and become a satellite and almost a colony" (129). Nehru was not unusual among educated Indians. In 1952, Dillon Ripley received an insightful letter from an Anglo-Indian who had decided to stay in India following its independence.

> A very high proportion of literate Indians, the kind who are vocal in politics, will argue with genuine conviction that the Russians and Chinese are today living under just as democratic a system of government as the Americans or the British. Moreover, to them the supreme issue in the world today is the issue of imperialism, not the issue of communism. Remember that it is only five years since the withdrawal of the British Raj. That is not long. They see the French still fighting to uphold an Empire in south-east Asia, and they know that the French get American help. They see American armies in Korea, and only one in a hundred will recognize that the U.N. armies are resisting aggression there. They tend to simplify the whole thing as meaning that the white west is still trying to encroach upon Asia. From this point of view, any further encroachment of American armies into any other part of eastern Asia would have a deplorable effect on opinion here, and might make the Government's task of trying to keep a sincere friendship with the West impossible. . . . Quite apart from possibilities of that kind . . . the ordinary Indian is very ready to believe that America is stepping into Britain's imperial shoes, so far as Asia is concerned.[5]

Quite simply, the United States was a bit scary to Indians who had just succeeded in getting rid of one group of English-speaking imperialists. The United States was powerful, it had global ambitions, and it expected all "right-thinking" nations of the world to fall in behind it. Its success in organizing a global coalition of armies against the North Koreans in 1950 only heightened this sense of unease.

When U.S. ecologists traveled to India to conduct research, they did so in the context of these uneasy Indo-U.S. relations. And just as many Indians were nervous about American geopolitical ambitions, many were also uneasy with Americans who came to India to do research—be they scientists or anthropologists. As Ali wrote to Ripley in 1945 with regard to their difficulty in getting permission to travel in Nepal, "They have a saying, which is unfortunately too true, that 'with the Bible comes the bayonet, with the merchant comes the musket.' What they fear coming with the ornithologist is not stated."[6] What "they" feared was given a more concrete outline in the late summer of 1950.

Once a Spy . . .

The August 26, 1950, issue of the *New Yorker* included an eighteen-page profile of Dillon Ripley.[7] The flattering account of the young (thirty-six-year-old) Yale professor's interest in birds and his world travels included a detailed account of Ripley's activities as the chief of the secret intelligence branch of the OSS in southeast Asia. It described how he trained and equipped Indonesian spies, all of whom were killed during the war. It described how he sent undercover men to Burma, Thailand, and Malaysia. It talked of his training, in India, among other places, and of his cooperation with the British military intelligence agents. But the author claimed that Ripley reversed the usual pattern—where spies pose as ornithologists in order to gain access to sensitive areas—using his position as an intelligence officer to go birding in restricted areas. The article undermined its own premise somewhat by also claiming that "Ripley's energy and tenacity . . . are usually concealed by a languid and lackadaisical air." It went on to say that many British intelligence officers underestimated Ripley and took him for an "eccentric country-squire type," when in fact Ripley was quite effective. As the head of the OSS, Maj. Gen. William J. Donovan said, "Ripley's previous contacts and experience, and his imagination, resourcefulness, energy, and tenacity, made him very useful to us" (32). After the war Thailand honored Ripley for his support of the Thai underground during the war.[8]

The *New Yorker* article confirmed the fears of those Indians who suspected that U.S. scientists such as Ripley—who kept requesting permission to walk around in sensitive border areas, like the forests along the border of India and China or Nepal—were collecting information that was not always strictly ornithological. After all, a man who had been a chief of secret intelligence in the organization that immediately after the war was reconstituted as the CIA was surely still keeping his eyes open and reporting back to his superiors in Washington. When the article was published, Ripley, as it turned out, was in the sensitive northeastern region of India, traveling near the borders of Bangladesh, Bhutan, Nepal, China, and Burma. His protests of innocence notwithstanding, he found forest after forest closed to his access, and he was shortly on a plane back to the United States.

Perhaps even more damaging to Ripley's immediate work in India than the revelation of his spy background was another element of the article. In his interview with the author, Ripley had told how he had finally gotten permission to collect birds in Nepal in 1947—a country completely closed to foreign scientists before then. He had pretended to be a close acquaintance and confidant of Nehru (whom he knew but was certainly not close to) when dealing with Nepalese officials. Ripley had implied that Nehru would be personally happy if Ripley could collect birds: "Mr. Nehru regarded this enterprise as a valuable one, and . . . it might carry a certain political weight" (34). The Nepalese government, anxious to start off on good terms with the newly independent India, gave Ripley the permissions he desired. The article then went on to imply that Ripley had played a key role in a new commercial treaty signed between India and Nepal: "Ripley is inclined to dismiss it as a fortunate non sequitur" (36)—but the implication was that Ripley was being modest. Nehru's fury is easily imagined.

In one fell swoop Ripley had alienated the new prime minister of India, an acquaintance of Ali's who was otherwise quite well disposed toward ornithology, and had confirmed the worst fears of those Indians suspicious of U.S. neo-imperialism. Ali was furious as well—his chosen collaborator on his planned-for masterpiece was now public enemy number one among the Indian elite. Ali wrote Ripley two letters. In the first he scolded Ripley at length:

> Surely no government can knowingly be expected to be enthusiastic over giving facilities for free movement in their jurisdiction to one who combines (or has combined in the past and may still be capable of combining) ornithology with secret service work. . . . [G]overnments are nervous creatures, and sometimes inclined to what may seem over caution once they smell a rat. Also I feel that the reference to your exploiting the supposed great friendship with Nehru in order to get admission into interior Nepal

may perhaps go down as clever with American readers, but is not in the best taste nor likely to be considered so either by Nehru or the Nepal authorities concerned. . . . I fear it may not redound to your esteem in scientific circles nor improve your position with the authorities in India. In fact many of the details could well have waited for your obituary notice! . . . I would gladly have spared myself the labor of writing had I been indifferent to your reputation both for your own sake and for the sake of our proposed future collaboration.[9]

Ali closed with "Yours sincerely," not his usual more casual "Yours." A couple of months later Ali was no longer so angry and now wrote to help Ripley repair the damage:

I have just heard from Horace the disturbing news of the Prime Minister's displeasure at certain parts of the New Yorker profile. This is what I greatly feared, and I am bound to repeat that I think it was most indiscreet on your part to have given such a very detailed interview to the paper. However, the damage has already been done which is a great pity. I hope you will lose no time in making a bee-line for New Delhi, and will get an opportunity of explaining things to J. N.'s satisfaction. It would not be so good if this came in the way of your further visits to India, and of your unrestricted movement in remote parts of the country.[10]

Ripley must have succeeded in his overtures to Nehru, for by January 1951, Ali wrote, "J. N. has been pacified and . . . the unpleasant episode will not interfere with your future visits here."[11] Nehru was forgiving, but Ripley's past as an OSS spy did not go away. His background fueled suspicions that U.S. scientists in India were actually, or also, spies. In fairness, these fears were sometimes well grounded. David Challinor, a former administrator at the Smithsonian under Ripley, commented that one of the scientists supported by the Smithsonian in South America, Wymberley Coerr, was later found to be "involved with Uruguay and the CIA."[12] Challinor also remembered a young man whom the Smithsonian helped fly to India, supposedly to do anthropological research, who unknown to them was also using a grant from the Department of Defense to interview refugees from communist China-occupied Tibet.[13] There were CIA spies in India, and at least one of them posed as a scholar—but there is no evidence that Ripley ever worked for the CIA or ever served as a spy after he left the OSS in 1945. Such suspicions are as difficult to disprove as they are to prove, however, and Ripley (and anyone he collaborated with—most notably Ali) would be plagued by these suspicions for the rest of his career in India.

Bird Banding at Bharatpur

Seven years after the stir over Ripley's *New Yorker* profile, Sálim Ali started down a path that would end with his own troubles with the Indian government and accusations that he was contributing to a U.S. biological warfare program. Ali's mistake was innocent enough. He wanted to band birds. Through the fifties Ali began organizing his field notes in preparation for the handbook that he eventually wrote with Ripley. As he collated his ecology and life history data, he noticed that his notes were sadly lacking for India's migratory birds.

One of the difficulties in describing the life history of a bird is that many species take off and fly thousands of kilometers at some point in the year. Ornithologists were thus able to describe only half the life cycle of migratory birds. Knowing that a bird flew south or that ducks of the same species were often found in a specific location in December did not suffice. Nobody knew if the ducks found in Ontario during the summer were the same as those found in Florida in the winter. Bird banding was the solution to this problem. By placing small metal tags on the legs of birds and then attempting to recapture them, scientists could monitor the whereabouts of individual birds. Banding was most successful for migratory game species. Hunters were relied upon to mail tags found on shot birds to the address on the tag, thus providing a (terminal) capture location. Bird banding had occurred informally in England as early as 1853, and in Denmark there are records of bird-banding in 1890.[14] Large-scale organized banding projects began in the twentieth century. In the United States, for example, the first bird-banding organization was formed in 1909 and the U.S. and Canadian governments jointly began formal record keeping in 1920.[15]

Banding was first practiced in India on a very small scale as a result of collaboration between the BNHS and a local sportsman–ruler. One of the maharajas of central India funded a BNHS bird-banding project in his state between 1924 and 1926 with metal tags that he had commissioned the BNHS to make.[16] The labor-intensive process (the band was stamped with an address and serial number, then cut and smoothed—all by hand) and the BNHS's lack of funds to hire a larger staff limited the number of birds that could be banded. The results were exciting however, with "recoveries reported from such distant places as Turkestan [*sic*], Siberia, and other parts of the USSR."[17]

The BNHS, on the basis of these results, encouraged other wealthy sportsmen in India to band ducks, but the limited supply of metal bands kept this from ever being a large-scale undertaking. When he had studied in Germany in 1929, Ali had witnessed a large bird-banding operation and longed to lead such a project in India,[18] but he could not get the necessary funds. Every winter when Ali traveled to Bharatpur he would paddle his small boat through the marshes, capturing and banding a few birds and dreaming of the day when he would find funds for a larger-scale bird-banding project.

Banding Birds for Medicine

Ali's chance came unexpectedly. In the mid-fifties an outbreak of a deadly new virus affected monkeys and villagers living near deep forests in the southern state of Karnataka. Today the virus is still considered highly dangerous—a biosafety level 4 pathogen.[19] Telford Work, an American scientist working in India for the Rockefeller Foundation, isolated and classified the virus, which was found to be a tick-borne hemorrhagic form of encephalitis (hence classified as a flaviviridae virus). The new disease was named the Kyasanur Forest disease (KFD).[20] The Virus Research Centre of Pune, India, noted that KFD was very similar to Omsk hemorrhagic fever and the Russian spring-summer encephalitis complex. From this, scientists in India and in the World Health Organization (WHO) began to wonder if the virus could have reached India carried in ticks on migratory birds flying south from the Soviet Union.[21]

This was not an unreasonable idea. During the summer of 1999, U.S. newspapers were filled with reports of the West Nile virus killing people in New York City. This virus had never before been reported in the Western Hemisphere. The U.S.-based Centers for Disease Control and Prevention (CDC) reported that the virus is not transmitted directly between people but instead is spread from birds to mosquitoes to humans. The virus has been found in crows, exotic pet birds, and other domestic birds along the East Coast of the United States; it is thought by the CDC that the virus spread to the United States via infected birds.[22]

From the little bit of bird-banding data that had already been collected, it was known that of the over three hundred species that wintered in India many migrated from the Soviet Union. Some species were easier to figure out than others, such as the aptly named Siberian crane. With an in-depth study, banding thousands of birds and testing the banded birds to see if they were carrying parasites or the virus, scientists thought that it might be possible to learn more about KFD. The WHO invited Ali to Geneva in March 1959 to attend a meeting of the Scientific Group on Research on Birds as Disseminators of Arthropod-Borne Viruses. As Ali wrote Ripley, "My chief interest in this subject lies in the possibility of our being able to wangle the WHO into financing the setting up of a bird migration study centre . . . a dream I have had for a long time."[23] As the birds were banded, the researchers would take blood samples and remove ticks. The ticks and blood samples could then be analyzed at the Virus Research Centre in Pune for the presence or absence of KFD.[24] Ripley confirmed to Ali that this was a good possibility—in fact, Ripley said, "We have a group here at the Medical School working on encephalomyelitis and its transmission," and Ripley had some of his graduate students working on birds with this project.[25] Ali's idea of combining bird studies with public health was not unusual; it was an established type of ecological research in the fifties and sixties (and to this day, with the West Nile virus). Ali's proposal was "warmly approved" by the WHO, who suggested that Ali also send samples to the Institute of Diseases with Natural Foci in Omsk, and the Institute of Poliomyelitis and Virus Encephalitis in Moscow. There, Soviet scientists could compare the bird samples with their own data on the two forms of Russian encephalitis to which it was thought KFD might be linked.[26] Ali agreed, and the BNHS Bird Migration Project was begun in 1960.[27]

Although Ali certainly cared about the fate of disease-stricken villagers, it is clear that he, like Schaller and Spillett in the sixties, was only too happy to use public health concerns to fund his ecological research. From a distance of nearly fifty years, it is sometimes difficult to remember that ecologists in the fifties, especially in the developing world, had great difficulty getting funding for their research. Today ecologists enjoy much higher visibility and have a much easier time getting funding, in large part due to the widespread awareness of conservation issues that permeates the world.[28] It is possible today for an ecologist to justify a research project solely on the grounds of gaining new information on a rare or understudied creature.

In 1959, though, the word *biodiversity* had not been invented. The Society for Conservation Biology was still twenty-seven years in the future. Rachel Carson had not published *Silent Spring,* Paul Ehrlich had not published *The Population Bomb,* and it was still legal to hunt tigers in India, for the right price. There was no Endangered Species Act in the United States. There was not much, if any, public awareness of an imminent environmental crisis, anywhere in the world. It was still two years before the World Wildlife Fund was constituted to combat the near impossibility of getting funding for ecological research and conservation projects.

Public health research had no such problems. People and funding agencies have always been relatively sensitive to human death and disease. Following World War II and the creation of the WHO, there was a series of global public health initiatives of shocking ambition, including programs for the eradication of malaria (1955) and smallpox (1966). These programs spent huge amounts of money; between 1955 and 1969, 1.5 billion dollars were spent on malaria eradication alone.[29] As with the new KFD virus in India, the WHO quickly mobilized funds for smaller projects as well. Of course, what seemed small to the WHO was huge for Sálim Ali. With WHO funds, Ali was able to order as many metal bands as he liked, of varying sizes, from Sweden. He could order mist nets from Japan to capture smaller ground-dwelling birds. And he could hire people—trackers and hunters from the Sahni and Mirshikar tribes, and young, otherwise unemployed ecologists from the BNHS—to carry out the intensive work required for this process.[30] Ali and his army of bird banders meticulously took blood samples and plucked ticks off of captured birds, and in exchange for this small chore the birds of Bharatpur were banded.

New Sources of Funding

To be effective, bird-banding studies must last for many years. It is not possible to band all the birds in a given locale in one or even five years: there are too many birds, capturing methods are too imperfect, and new birds are born every year. Further, a banded bird in and of itself yields little data other than size and approximate age (or infection level for a public health study) at time of capture. The true power of bird banding as a technique is derived from the recapture of these birds in successive years. Birds can be recaptured in the same place in later years, indicating that this species might return to the same winter or summer site year after year. Recaptured birds give an indication of life spans and growth patterns. Birds recaptured far from the place they were banded show migration patterns. The recapture of a banded bird, though, is very difficult. In actuality, most banded birds are never seen again. In order to get a reliable (meaning large) data set, huge numbers of birds must be banded.[31]

The WHO did not have the same commitment to long-term bird banding as did Ali and the BNHS. The KFD virus did not develop into a crisis in the sixties.[32] And while the BNHS Bird Migration Project was yielding a great deal of ecological data on bird migrations and lifecycles, it did not seem to have direct public health import. The KFD virus was not spreading and the sense of crisis had passed. As the United States cut back on its donations to the WHO in the mid-sixties (in part to fund the Vietnam War), the WHO decided to stop funding the BNHS project.[33] Without this external funding, the bird-banding project could not continue. The WHO and Ali's international connections stood him in good stead, however. His friend and collaborator Dillon Ripley was the head of the Smithsonian Institution by this time, and the Smithsonian "considerately took over the funding for the interim period of uncertainty."[34] The Smithsonian did not have to help for long—almost immediately, the WHO arranged for another source of funding for Ali, the Migratory Animals Pathological Survey (MAPS).[35]

MAPS, another large-scale bird-banding project in Asia in the sixties and seventies, dwarfed the BNHS project. Birds were banded throughout Southeast Asia, including Japan, the Philippines, Thailand, Hong Kong, South Korea, Malaysia (then Malaya), Indonesia, and Taiwan.[36] In its twelve years of existence (1963–75), teams of ornithologists banded over 1.2 million birds, of over a thousand species. Blood samples and parasites were also collected from the birds and analyzed for viruses, primarily Japanese encephalitis. There were only fifty-five hundred recaptures or band returns, (half of one percent of the birds banded).[37]

Elliott McClure, a bird bander par excellence who was credited with personally tagging more than a hundred thousand birds from five hundred species in his lifetime, led this staggering program.[38] He received a Ph.D. in wildlife management from Iowa State in 1941, and after a stint in the navy during World War II (working with public health in California) he had been hired by the army (as a civilian) to investigate encephalitis in humans, birds, and other animals. McClure had spent seven years in Japan and five years in Malaysia doing this work, very similar to MAPS but on a much smaller scale, prior to 1963 (102–79). MAPS was conceived by the U.S. Army Medical Research Laboratory, and in 1963 McClure was approached by air force colonel Charles Barnes and asked to head the project. The U.S. Army Research and Development Group had agreed to fund an international investigation of migratory bird patterns throughout Southeast Asia, hoping to better understand the regular spread and retreat of Japanese encephalitis across the Asian landmass (280–81). McClure readily agreed to head the program, and he promptly began putting together the national teams who would do most of the work on the ground. McClure's teams had just begun banding birds (the program was running by 1964), when the WHO decided to stop funding the BNHS bird-banding project in Bharatpur. McClure had been following the results of the BNHS project, which was so similar to his own, and as Ali wrote, "on learning of the possibility of its having to be wound up promptly offered to step in with the funding on condition that our data and results would be made available to [the MAPS program] to complete its own information."[39] Ali readily agreed, and the banding at Bharatpur continued. But now Ali's bird-banding project was funded by the U.S. Army Research and Development Group!

PL 480 and the Smithsonian Foreign Currency Program

In 1966 the Smithsonian Institution succeeded in receiving access to a new source of funds, to be used for international research. Under Public Law 480 (Food for Freedom), in effect from the mid-fifties through the sixties, the U.S. government had decided to sell agricultural goods to developing nations rather than give them away. The government had allowed these nations to pay the United States in local currencies, however, which were held by their governments. Thus, no money actually had changed hands. By the mid-sixties the U.S. government technically held staggering amounts of local currencies for some nations, such as Poland, Egypt, Pakistan, and Guinea. In 1967, India had approximately $630 million worth of rupees—the highest dollar amount of any nation.[40] These rupees did not exist except on paper. By the early sixties the U.S. Congress began demanding that these "surplus funds" be used. The other nations involved fought this vigorously. If Indian rupees were used (instead of dollars) to purchase Indian mangoes for a U.S. market, for instance, India would not only lose dollar income, but would also in effect be using its own revenues to provide the United States with goods. The author of a commissioned report on the issue stated, "A citizen of India assured me that his nation would fight most vigorously against the use of U.S.-owned rupees. When asked why India agreed to pay in rupees in the first place the reply was, 'They are only paper to us,' and he was quite sure his government could postpone their use indefinitely" (5–6). Recipients of PL 480 money hoped to never have to pay for the agricultural goods. This became a source of contention during the sixties, especially as the United States suffered a balance-of-payments deficit.

By 1965 a compromise solution was in place. The foreign currency holdings would be available to U.S. researchers in those countries, particularly those in archaeology and related disciplines, like linguistics and art history. In this way, the foreign currencies and the goods they purchased never left the country, and perhaps most important, the foreign currencies did not replace U.S. dollar expenditures—without the foreign currency funds, the research simply would not have taken place. The Smithsonian "was perceived as a relatively neutral non-federal agency who would be willing to evaluate proposals to work in these countries in a fair fashion," so it set up review committees and served as the grant clearinghouse for this program.[41] Almost immediately, Ripley (now secretary of the Smithsonian) began advocating the use of the PL 480 money for ecological research, "especially [in] India, which I consider the most promising."[42] Ripley was successful, as the Department of the Interior and Related Agencies Appropriations Act, 1966, expanded the valid use of PL 480 funds to include the natural sciences and cultural history. With access to hundreds of millions of dollars, for the next few years it seemed that the sky was the limit for imagining ecological science projects in India.

Ripley and Ali were obviously well situated to benefit from the new source of funds. As soon as it was clear that money would be available for the natural sciences, the director of the Smithsonian Foreign Currency Program wrote to Ali that it "had been brought to my attention" (by Ripley, no doubt) that Ali hoped to expand his bird-banding program into new regions of India and that "this work might be eligible for support in the form of a grant in Indian rupees under the . . . Program. If you would care to submit a research proposal with a budget of perhaps up to $10,000, we would make every effort to obtain prompt review by our scientific review committee or to obtain other financial support to enable you to extend your excellent work."[43] By the end of 1967, Ali's bird-banding project was flush with funding. The Smithsonian funded Ali's work on migratory birds at the Bharatpur field station from September to March (Soviet zoologists joined them at the Bharatpur field station in 1968). The MAPS program (and the U.S. Army) then funded Ali's work on birds in the western state of Gujarat from March to September, as well as less-organized bird-banding activities in eastern India throughout the year. Ectoparasites and blood smears were being sent to MAPS, mallophaga (biting lice) were being sent to the Smithsonian, virological studies were being conducted at Pune, Moscow, and Omsk, and bird-banding records were being copied and stored at the BNHS, the Smithsonian, and the MAPS office.[44] At the same time, PL 480 money was helping to pay for the publication of Ali and Ripley's *Handbook,* the first volume of which was released in 1968.

Elsewhere, Smithsonian Foreign Currency funds were paying for ecological projects throughout India, most notably on the ecology of the Gir lions in Gujarat. As 1969 began, India was preparing to host the world meeting of the International Union for the Conservation of Nature in New Delhi (supported by rupee funds), a national chapter of the World Wildlife Fund was being set up in India, and talk was growing of the need to do something about tigers. The Smithsonian was preparing a proposal for three permanent research stations in India (one for the Gir lions, one for the tigers at Corbett National Park, and one on the rhinos at Kazaringa National Park). This was the high point of Smithsonian-BNHS collaboration in India. After 1969 neither institution would again wield such easy and apparently limitless influence on the ecological sciences in India. In 1969 politics again caught up with the ecologists and first Ripley and the Smithsonian, then Ali and the BNHS, had to answer some difficult questions about their science and their U.S. military funding.

The Pacific Ocean Biological Survey

On February 4, 1969, at 9:00 P.M. Eastern time, the NBC television network news show *First Tuesday* presented an episode entitled "Chemical-Biological Warfare (CBW): The Secrets of Secrecy." It opened with a videotape of a rabbit being exposed in a laboratory setting to nerve gas. The screen showed the rabbit convulsing, then dying, as the voiceover explained, "the United States also has nerve gas for people." A substantial section of the show was devoted to "an ultra-secret test project in the Pacific Ocean, conducted under a cover of bird-banding study [*sic*]." The narrator continued: "The Pacific Ocean biological survey has cost the Defense Department more than two and a half million dollars. This amount was paid over the past six years to the Smithsonian Institution."[45] That same day, the front page of the *Washington (D.C.) Evening Star,* reported, "The Associated Press quoted NBC correspondent Tom Petit as saying . . . that the Pacific project 'in effect was a [test] of an animal delivery system for chemical-biological warfare.'"[46]

The television show, and its associated press coverage, was sensationalistic, but in its broad outline it was correct. Beginning with the army approaching the Smithsonian in 1962, the Department of Defense (DOD) had funded a massive bird-banding and ecology study in the central

Pacific. This study, POBS, was not linked to the MAPS study—they were funded through different branches of the army, and there is evidence that they were not coordinated.[47] POBS was not a cover, as subsequent reports made clear (including a detailed review of the project in *Science*).[48] Instead, the Smithsonian study was exactly what it was purported to be—an attempt to determine what diseases birds of the central Pacific naturally carried and to discover bird migration patterns in that region. It is also clear that POBS was connected to the U.S. biological warfare program. The army claimed that it was only interested in the study because it hoped to better understand both why its soldiers got sick in the Pacific and how to avoid hitting birds with airplanes on small island landing strips; but POBS was clearly more than this. The study was run out of Fort Detrick, Maryland, the army's biological warfare center, and blood samples, ticks, birds, and data were regularly sent both there and to the Deseret Test Center in Utah, a testing site for chemical and biological warfare agents. Scientists who worked on the project were given security clearances, some of their documents and research results were classified, and some were given inoculations before being sent to the field to collect data.

Other researchers have since determined that the army wanted to conduct biological agent tests. They wanted a test site that mimicked the climate of the Asian regions in which the United States found itself currently fighting. They needed an area that was remote, and they needed to make sure no birds would fly by the island, stop for a rest, pick up a lethal disease, and then carry it to human population centers. Ideally, for their test the army wanted to introduce only diseases (or close variants) that were already found in that region. The Smithsonian POBS study satisfied all these requirements. It would tell them where birds migrated. It surveyed the flora and fauna of every island in the central Pacific. And the researchers sent back to Maryland blood samples and insects that could be tested to see what diseases naturally occurred among the bird populations. And the army did conduct at least one test—though not with a live biological agent, it appears—in 1965.[49]

In this way, studying migratory birds could be presented as the acquisition of knowledge to help the U.S. defend itself from biological attacks or even to know when a biological attack was occurring. In a 1999 NATO conference on biological warfare, ecological studies of animal epidemiology were encouraged.[50] Without a solid understanding of how diseases spread via animal vectors (including birds), it is impossible to know when a disease has been artificially introduced. As the author of the briefing paper mentioned, new diseases emerge every year. Local populations often blame these diseases on hostile governments, as when the U.S. government was blamed in some parts of the world for the introduction of the AIDS virus in the eighties. Solid epidemiological work can help counteract such claims. Similarly, if a human-introduced disease were to arise, epidemiological work could help determine how it was introduced and from where it might have come. Massive bird-banding projects on migratory birds help fulfill this goal by determining where birds go and what diseases they might normally carry. This was how McClure presented the MAPS bird-banding project in Southeast and South Asia in his autobiography.[51]

Studying the transmission of biological pathogens by birds for defensive purposes is only a hair's breadth from turning that information to an offensive purpose. Some journalists claimed the army wanted to determine how to use birds as carriers of biological diseases (rather than dropping the pathogens from planes or other obvious or risky delivery system). The idea is not entirely ludicrous. The Smithsonian's assistant secretary for science at the time of the POBS scandal, Sidney Galler, was reported as believing in that possibility, claiming that some Pacific birds can "migrate tremendous distances and reach target areas with about 97 percent accuracies."[52]

The U.S. Navy conducts some of the world's most sophisticated research on marine mammals, particularly dolphins. Their Marine Mammal Research program began in 1959. It is not oriented toward abstract knowledge. Dolphins were studied to see what use they might have in marine warfare—could they patrol harbors and attack enemy scuba divers who might try to sabotage ships? Could they locate underwater mines? Pigeons were a staple of warfare, carrying messages before modern communications made their wings obsolete. After that, pigeons were used experimentally as potential guidance systems for long-range weapons (the pigeons would carry a small transmitter). Armies have always been more than willing to enlist nonhuman help whenever appropriate, be it Hannibal's elephants, German shepherds, or even migratory birds.

The U.S. military was not in a morally strong position during the latter stages of the Vietnam War. Many U.S. citizens, let alone people in the rest of the world, were only too willing to believe in the most nefarious interpretation of military actions, as well as the government as a whole. The world had been in turmoil in 1968, and the United States had seen riots and the assassinations of Martin Luther King, Jr., and Robert Kennedy. Napalm and Agent Orange were being dropped on the jungles and villages of Southeast Asia. The My Lai massacre was front-page news throughout the world. During this period even nonmilitary medical personnel in the U.S. seemed to have lost their moral compass and some were willing to carry out experiments with little regard for human suffering—the Tuskegee syphilis experiment, in which treatment was withheld for over forty years from a group of African-American men in Alabama who thought they were being treated, was publicly exposed in 1974.[53] There were a number of less visible operations at this time as well, particularly those involving the CIA, which border on the bizarre. The CIA was linked to the introduction of swine virus into Cuba in 1971, which ultimately killed half a million pigs.[54] And CIA attempts to assassinate Castro are legendary (including experiments with exploding cigars and poisoned scuba suits). Through the fifties and sixties the chemical and biological warfare unit of the U.S. Army conducted several well-documented tests to see how rapidly an introduced (benign) bacteria would spread in a city, using the unwitting citizens of San Francisco, New York City, and Minneapolis, among others. As with the bird migration studies, such studies could be thought of as offensive (how could we attack the Moscow subway system?) but they were presented, at least, as studies oriented toward gaining knowledge about how vulnerable the United States was to terrorist attacks.[55] As we consider the POBS project, absurdity, or a belief that the U.S. military or government would not do such a thing, are not sufficient bases to dismiss claims that the army considered infecting birds with deadly viruses. During the Cold War the U.S. intelligence community was a strange batch of warriors indeed.

In the inferno of press coverage about the Smithsonian link with POBS and biological warfare immediately following NBC's 1969 report (including articles in nearly every major U.S. newspaper), many of the Smithsonian's leaders claimed they had not been aware of the full nature of the POBS project. In statements immediately after the report, the assistant secretary for science categorically denied that the Smithsonian study was helping the army find a test site for biological agents.[56] When NBC had been investigating POBS for their report, a Smithsonian researcher answering their questions wrote, "The Smithsonian Institution does not conduct classified research for the Department of the Army." That was incorrect—some of the POBS work was originally classified. Later in the letter he said, "Fort Detrick has no special interest in tropical ornithology, and as has been explained above, there is no one at that agency who is intimate with the Smithsonian projects."[57] Again, that is incorrect. Surely he would not have been so direct had he known the truth—he would have covered himself at least a little. Similarly, at that time other leaders from the Smithsonian denied knowing of any classified work conducted by the Smithsonian. The sheer number of Smithsonian officials who later claimed to have learned the full scale of details about the project only in the eighties suggests that the "secrets" of POBS were truly secrets, even from many Smithsonian officials who knew that the DOD was funding this massive bird-banding study but did not know its link to Fort Detrick and biological warfare.

In late 1969, Ripley wrote Sálim Ali, "The fact that we applied for a grant from the Defense Department for work on migratory birds in the eastern Pacific, fully in continuation of the principles of the Whitney expedition and in total innocence of any underhand project, has now been established."[58] Ripley had no reason to lie to Ali—and in fact their correspondence is replete with instances of honesty that can make an outside reader cringe (evaluations of common acquaintances, of governments, or the like). In fact, Ripley had every reason to be honest with Ali, as he and Ali had often dealt with similar questions of the propriety of their work. It appears that Ripley, even after the scandal broke, did not know the full extent of the POBS projects. He believed that the Smithsonian had approached the DOD (when in fact the reverse was true), and he did not think that the military had any goals in supporting the project. That had been the official Smithsonian line, and it appears that those scientists who knew better (who directly worked on the project) kept quiet for several years. Although all the records at the Smithsonian are now open, with any formerly classified project there is always the possibility that the U.S. Army kept some records that were not made available to the Smithsonian. At least one Smithsonian scientist involved with the project, Charles Ely, regularly burned carbons and records of his communications.[59]

The controversy over POBS had far-ranging effects. The U.S. Congress questioned the army's Chemical and Biological Warfare program, and began asking difficult questions about why biological warfare research and tests should be secret when nuclear-war research was not. The Smithsonian's 1969 annual report states, "a decision was made at the Smithsonian . . . as a direct result of the publicity generated by NBC's . . . TV program, to terminate field work except on Sand/Johnston [Islands] by 30 June, 1969."[60] By June 1970 even the last remnants of the POBS program had been shut down. The Smithsonian vowed to never again allow any of its research to be classified. More dramatically, by that time, the entire U.S.

biological warfare program had been shut down. On November 25, 1969, President Nixon ordered an end to any offensive biological warfare research or testing, and by February 1970 he had extended that ban to toxic agents of any sort.[61]

The POBS study opened up a discussion in the United States about the propriety of academic and research institutions accepting funding from the DOD. Was it possible to accept such funding without accepting that it was war related? This question was particularly pressing in 1969, at the height of the Vietnam War and the civil protests against it. The Senate wrote the so-called Mansfield amendment, which explicitly stated that "projects receiving Defense Department support are indeed of military importance and can no longer be considered blameless research." The Smithsonian promptly began checking that none of the research projects in their Foreign Currency Program were jointly funded by DOD grants. They were determined to insure that the Smithsonian would not again be implicated in questions of secret defense-related research. Three Smithsonian projects fell in this category—a bird study in Egypt, a mammal project in Morocco, and Sálim Ali's bird-banding study in India (which was still receiving money from MAPS, and thus from the U.S. Army). In an official memorandum, the Smithsonian urged the leaders of the Egypt and Morocco studies to disassociate themselves from the DOD, but was strangely silent on Ali, except to say "the Indian Bird Banding project, nevertheless remains vulnerable as Dr. Sálim Ali is still receiving MAPS money." It seems likely that no official memorandum was needed for Ali, and that he and Ripley had already been working to disassociate the Indian project from the military, for the POBS scandal had spread to India.[62]

Questionable Funding Sources Come Home to Roost

The POBS scandal reached India almost immediately after the NBC telecast. Within a week, the Smithsonian was advising the U.S. ambassador to India (at his request) of Ripley's involvement and funding sources in India. The scandal in India was threefold: Smithsonian involvement in the Bharatpur bird-banding study, a rumored radar system that was supposed to track birds (but really would be useful militarily), and trips by Ali and Ripley to the small Himalayan kingdom of Bhutan. A portion of the NBC report that had been ignored in the United States—a throwaway line that "A scientist in California has been asked to develop a bird-counting radar," spurred the radar rumor. It was easily dismissed, as no such plans appeared to be in place—perhaps people were confusing that with a DOD-sponsored study in Calcutta of the ionosphere. The other two concerns were more relevant—obviously, the Smithsonian and MAPS were funding the Bharatpur study, and Ali and Ripley had made trips to Bhutan (and planned to do more there). Ali moved quickly to differentiate his Bharatpur study from both the POBS project and more direct collaboration with Ripley (as in Bhutan). He apparently was successful. At the end of February, Ripley wrote Ali, "I am happy that you have been able to establish that your work in Bharatpur is distinct from the work that has been undertaken by us both jointly in Bhutan." The Bhutan accusations were in some ways the most serious, at first. Bhutan was very much a monarchy and Ali and Ripley's work in that country depended on the personal goodwill of the king. Ripley was a threatening figure because of his previous background in the OSS supporting underground resistance movements and he was accused of "somehow attempting to sponsor independence activities in Bhutan!"[63]

As 1969 stretched on the POBS scandal died down in the United States, but in India accusations of U.S. spying and intrigues via bird banding continued. Ripley and Ali's correspondence reflected a bit of surprise at the strength of the Indian response, months later: "I too am deeply distressed and disappointed by the continuation of these scurrilous and unfounded rumors about our supposed Defense-related research in India. . . . I am deeply sorry about the whole thing because of course it provides continual food and ammunition to careless persons in Parliaments around the world or those who actually intend us some harm." Ali was a difficult target for those who wanted to paint his migratory bird studies as completely subject to U.S. military goals. Soviet scientists were intimately involved in the research and Ali made a trip to the Soviet Union in September to visit colleagues there. He reported to Ripley, "The Russians were most friendly and co-operative."[64]

At least formally, the debate about Ali's bird-banding studies focused more on the propriety of accepting U.S. Army funds than upon spying accusations. The large English-language daily the *Times of India,* reported in August on U.S. Senate debates about DOD funding of academic research. The article clarified that there were two such studies approved for 1970 in India—a radio-astronomical and satellite study of the ionosphere and a combined BNHS study of the bionomics and taxonomy of the birds of India and the taxonomy of the birds of Bhutan. It also relied on statements by Senator William Fulbright to make its main point about U.S. military funding of research in India: "He said that such activities injured Indo-American relations 'because we intrude with Defense Department funds into foreign academic research . . . what we are doing is driving friendly countries away from us. They do not wish to be an appendage of the Pentagon. I think they are quite justified.'" This moderate line seemed to prevail in India throughout the end of 1969 and 1970. The more radical critiques did not seem to be a threat. Ali began to disassociate himself and the BNHS from MAPS and moved to rely solely on PL 480 funds through the Smithsonian's Foreign Currency Program. Ali and Ripley hoped that the firestorm of 1969 would, like the controversy surrounding the 1950 *New Yorker* article, fade into the past. As Ali quoted to Ripley, "The dogs may bark but the caravan must go on." Unfortunately for Ali and Ripley, the caravan ran into a roadblock and the dogs ended up being quite persistent.[65]

The Gateway to India Is Closed

Throughout the various controversies of the fifties and sixties, Ali and Ripley could depend on a basically favorable climate in the government of India for their ecological work. First Nehru and then his daughter Indira Gandhi were supportive of Ali and the BNHS throughout the fifties and sixties. If a few members of Parliament or the press should attempt to make a stir about the BNHS-Smithsonian collaboration, Ali and Ripley would write their letters (to ambassadors, to prime ministers, to highly placed friends) and the storm would pass. In particularly sticky situations they would make personal visits and smooth the waters. Although the United States and India had disagreements at the level of international politics, each government was committed to not alienating the other. Kissinger described this relationship as "frustrated incomprehension within a framework of compatible objectives." This stood Ali and Ripley in good stead on a number of occasions—their scientific collaboration represented a type of cooperation between the two nations that both governments could support, regardless of other momentary differences. Thus, a government official could praise the Ali-Ripley work as indicative of the deep and abiding friendship of the two nations, and then turn around and critique the misguided (U.S. or Indian) policy on some thing or another. This all ended with the 1971 India-Pakistan War and the U.S. "tilt toward Pakistan."[66]

When India and Pakistan went to war in 1971 over the fate of Bangladesh, President Nixon chose to support Pakistan. He felt this was necessary on three counts: the U.S. had an alliance with Pakistan, Nixon was trying to arrange his visit to China (which also supported Pakistan), and India had just signed a friendship treaty with the Soviet Union. Both the U.S. State Department and the CIA opposed Nixon's decision to support Pakistan (their analysis of both India's and Egypt's friendship treaty with the Soviet Union was that it merely allowed both nations to be flexible in their appeals for foreign aid) (1284). At the height of the brief war, Nixon ordered a U.S. naval fleet to sail into the Bay of Bengal as a measure of support for Pakistan. On December 6, 1971, the U.S. cut off economic aid to India—though Kissinger claimed that the State Department was so unenthusiastic about this policy that it was never effectively implemented (900). Nevertheless, 1971 was a tumultuous year for Indo-U.S. relations, and the end of that year saw a distinct chill settle over even cultural and scientific collaborations between the two nations.

Ali Accused

At the end of 1971 the government of India froze PL 480 funds. They refused to fund new projects and did not renew funding for old projects. This policy was not directly announced—it emerged in practice, as project proposals would enter a bureaucratic never-never land. In meetings government officials would state that the PL 480 funds were "under review." If the United States was going to eliminate aid to India, why should India continue to sponsor U.S. researchers in India? Ali and Ripley promptly began their usual routine. They wrote letters. They made visits to New Delhi. And they went all the way to the top. In October 1972, Ali wrote Prime Minister Indira Gandhi, describing the history of his bird-banding project. Very aware of the changed political climate, he fudged a few dates—in his summary for the prime minister, the MAPS program left India in 1967 and was replaced by the Smithsonian in 1968. The Smithsonian had indeed begun funding the bird-banding program in 1968, but in conjunction with continued MAPS funding—at least through 1970. Ali also emphasized his collaboration with the Soviets in this letter, devoting much space to describing the various labs in the Soviet Union working on the viruses, the visits of Soviet scientists to India, and his visit to the Soviet Union: "The collaboration of the Russian virologists . . . has been maintained over the years and has proved highly beneficial to both parties." He did not mention if U.S. scientists ever visited the study sites. Ripley is acknowledged as a co-investigator only at the bottom of the second page of the letter. Ali went on to explain that, although the BNHS had been approved for a PL 480 grant, the Ministry of Finance refused to release the funds and that "all our bird migration activities have had to be suspended, and unless the situation can be eased without delay we may be obliged to wind up the project."[67]

Ali's appeal succeeded, but not in the way that he had anticipated. The government of India stepped in and gave Ali funding from its own sources, not from the frozen PL 480 accounts. This respite from funding woes lasted just over a year. In 1974 an independent Indian journalist described Ali's involvement with the U.S. Army's MAPS program (and claimed that it was ongoing). The journalist suggested that, in Ali's words, "the Society was in this way colluding with the United States to explore the possibility of migratory birds being used in biological warfare for inducting and disseminating deadly viruses and germs in enemy countries." This was not a new idea, but it was expressed in a different political climate. This report caused an uproar in the Indian parliament. In May, Ali wrote a nervous letter to Ripley, "One of our super-patriotic members of Parliament again tried to flog the dead horse of the BNHS having accepted funds from the U.S. Army (MAPS). . . . Since no awkward repercussions have taken place as yet in the way of our being asked to explain, I expect the charge was not taken by anyone seriously. However, it is not yet too late for such an official demand to come, so we are keeping the appropriate powder and shot ready. A plague on these evergreen suspecters of everybody and every activity!" As he feared, further parliamentary discussions followed, and in August Ali attempted to defend himself in a letter to the *Times of India*. The letter was published, but the suspicions did not simply go away this time.[68]

This scandal included its share of misinformation, including continued fears that the bird-banding project included radar in northeast India, to check on China—when in fact no bird banding had occurred in that part of India, let alone with radar. The 1974 criticisms also claimed that Ali was still involved in the MAPS program. While the MAPS program was still running in other Southeast Asian countries (it ended in 1975), it was no longer funding work in India. Where these critiques were more damaging to Ali, however, was in focusing more directly on the implications of the MAPS project itself, regardless of when it had been in India. Earlier protests had discussed the impropriety of DOD funding in India or Ripley's OSS connections but had not looked so directly at the specific project involved.

U.S. military involvement in the Bharatpur bird-banding study could appear suspicious to even a neutral observer. The original intent of the study was to track the spread of the KFD virus and determine if birds (naturally, not via biological warfare) carried the Virus from the Soviet Union to southern India. In the process of disproving this theory, it was necessary to ascertain whether viruses could be transmitted from birds to people in such a fashion, and how the virus, once introduced, would spread. Such information could as easily be applied to the intentional introduction of a virus as it could be to tracking the etiology of an already existing virus. And through careful migration studies, scientists could predict exactly where a bird would spend its summers, and thus where the introduced virus might spread.

There had already been research on the effect of the KFD virus when introduced to people; a study recorded in the *British Medical Journal* had linked tests using the KFD virus with two scientists from the official site for British biological warfare studies at Porton Down.[69] Australia lists the KFD virus on its biological weapons warning list.[70] And MAPS was not merely a scientific undertaking, it was a program run by the U.S. Army Medical Research Laboratory and SEATO, funded by the U.S. Army Research and Development Group, and implemented when the Vietnam War was at its peak.[71] In addition to KFD, this bird-banding research on migration could be applied to introducing any one of a number of tick-borne diseases. The resulting images are powerful: flocks of geese flying overhead, an avian air force, unwitting carriers of lethal viruses in the ticks on their bodies, the ducks in a Moscow city park infecting the very children that fed them crumbs. Hollywood would be hard pressed to do better.

Fears of biological warfare in Asia were not new and extended beyond India and beyond the 1969 POBS scandal. During the Korean War the Chinese and North Korean governments presented evidence that the United States was dropping biological warfare agents out of airplanes. The Soviet Union supported these claims, after sending a team of experts to review the data. It was an international scandal. The United States proclaimed its innocence, but proving innocence in such a case is difficult. Many people in Asia and throughout the world believed that the United States had dropped biological agents on North Korea. The United States was not cleared of these charges until 1998, with the discovery of some documents in a Russian presidential archive that proved that the Soviet Union had helped to manufacture these false reports.[72]

The controversy over Ali's project was an extension into India of longstanding protests, particularly by the Chinese, over supposed U.S. biological warfare studies involving bird migration and viruses in Asia. Elliott McClure, the leader of MAPS, was banned from travel in China or the Soviet Union: "I had been on the [Communist hit] list since handbills were distributed in Tokyo in 1950 describing our work with birds and distorting it to be a study in biological warfare. We were supposed to be inoculating birds with viruses and freeing them to take infection to China and other lands." China absolutely refused to collaborate with the MAPS study, and no bird bands were ever returned to MAPS from that country. The Soviet Union was more moderate and, as we have already seen, many scientists actively contributed to the migration studies and virology research of both the BNHS study and MAPS.[73]

Faced with this new wave of criticism in 1974 and 1975, Ali and his supporters made the usual protests of innocence. The official stated goal of the U.S. Army for the MAPS program was to learn how a disease such as Japanese encephalitis was reintroduced to semitemperate Asian areas after it had receded for the winter. Solving this question might lead to measures that could improve the health of soldiers in the field (and, implicitly, improve the U.S. Army's efficiency in the jungles of Vietnam). Ali, of course, claimed he was only looking for funding for his bird-banding study, a research project with no extraecological motivation.

The declarations of innocence did not help. All funds for the bird-banding study were frozen, and the government of India formed a committee of inquiry. Although it cleared Ali and the BNHS of "criminal intent or subversive action," it did not appease Ali's critics, and a second committee was constituted. Its findings were the same. Still, though, there was widespread general suspicion that Ali had indeed been collaborating with the U.S. Army and that the BNHS project had been research with the goal of conducting biological warfare. A third committee, made up of senior scientists from three different government institutes, conducted yet another study of the Bharatpur project and once again the BNHS and Ali were cleared of any actual wrongdoing.[74] After two years of investigations—including no funds for bird banding—the government of India resumed its funding, and Ali's bird banding studies continued.[75]

Whatever the goals of the U.S. military, and although Ali was officially cleared of "criminal intent or subversive action," he could not so easily clear himself of having willingly assisted the U.S. military in acquiring information it desired.[76] In India, collaborating with the U.S. Army has never been seen as anything less than unpleasant, and especially so during the immensely unpopular Vietnam War or after the 1971 India-Pakistan War. In the end, there are many ways to try to understand Ali's decision to collaborate with the MAPS program. His friends might say his decision was not politically savvy. His enemies *did* say it was treasonous. Others, not sure, might see it as a Faustian pact, selling his soul for knowledge and a few more banded birds.

"A Spook under Every Wing"

The scandal over Ali's connections with the U.S. Army was only the most spectacular and far-reaching example of Indian fears that the U.S. government was attempting to use ecological research as a cover for clandestine activities. Smaller examples abound. In 1982 scientists from the U.S. National Zoo (part of the Smithsonian) were conducting research on the gharial—a fish-eating crocodile. These scientists were radio-tracking the gharial in the low-lying terrai region of Nepal, near the border of India, to determine their movement patterns and range requirements. This required, of course, that transmitters be attached to the animals. During the monsoon, one of the gharial floated downstream and into India. It was spotted in India by a fisherman, captured, and turned over to the authorities. Newspapers reported that the CIA was attempting to map India's border rivers by means of floating crocodiles sending radio signals to satellites. This story has reached folklore status in India.[77]

This chapter began with a newspaper quote from another "spy story." When a U.S. ecologist was flown from Nepal to India to tranquilize (via darting) a man-eating tiger, so that it could be relocated, some journalists saw a potential CIA plot. He was working in the sensitive Sunderbuns region, on the border of newly independent Bangladesh. When the tiger eventually died (after being darted and relocated) this just added fuel to the fire—clearly the ecologist was incompetent, the press thought, so his real expertise must be elsewhere, as in spying perhaps.[78] In short, the accusation that Indian ecologists are "agents of foreign powers out to sabotage India's forward march" is a common one, and one usually made with far less basis in fact than with the MAPS study or the fears about Ripley's past. Noted environmental historian Ram Guha and ecologist Madhav Gadgil suggest that such arguments are made not just by believers, but more cynically, by proponents of development projects who wish to disable the ethos of ecologists suggesting political alternatives to large projects like dams.[79]

Although many such accusations were unfounded, ecologists in India and elsewhere in the world acted throughout the fifties and sixties as if their research funding had no political or ethical import. They had been accustomed to taking money from any source and attempting to do their own work while fulfilling grant requirements (as with all the public health funding of ecological research). In doing so, they did not just open themselves to criticism by journalists and government officials. They implicitly supported institutions or policies that they might not have actually supported. When Ali and the BNHS decided to accept funding from the U.S. Army, they were supporting the U.S. military in its goals.

The BNHS-MAPS collaboration ultimately proved sensational enough that it drew the sustained attention of a good portion of the educated populace of India and the Indian government. The stereotype of the otherworldly ecologist who does obscure work with no tangible import was displaced by the idea of ecological research that was implicated in national pride and national security. Whether Indians thought Ali was guilty of spying or not, they were disturbed by the ease with which U.S. agencies like the army could fund research and collect data in India. In the context of frosty Indo-U.S. relations, the government bureaucracy in India decided that the funding and conduct of ecological research was too important to be left to the scientists, who had displayed a notable lack of self-control with regard to funding. Thus, in the early seventies the government began to control research projects in India more tightly, especially the PL 480 funds, and strictly limited foreign grants and funds from entering the country. The cost of Ali's bird-banding research was paid in the decreased freedom of ecologists in India and the United States to collaborate, in the loss of many of the collaborative funds that the BNHS had relied on to subsidize so much of their work raising environmental awareness in India, and in greater governmental control of scientific research in India's national parks—and by extension, greater control by the government of the information and knowledge used to manage 5 percent of India's landmass and millions of people. After 1975 the banded birds of Bharatpur continued to migrate with impunity, but not so U.S. ecologists and money.

CHAPTER FIVE

"Modern" Ecology Comes to India

Madhav Gadgil and the CES

> "Nothing in biology makes sense, except in the light of evolution." This asser-
> tion by evolutionary biologists has only recently percolated to ethology, the
> study of animal behavior. But the result has been spectacular. Many aspects of
> the behavior of animals, especially social animals, that appeared paradoxical
> or could be described but not fully understood, now appear to have a logic.
> For the first time we are truly able to ask why an animal does what it does.
>
> —*Raghavendra Gadagkar*, **Survival Strategies**

ON A MID-DECEMBER DAY in 1965 two men found themselves alone on an elevator together—a professor and a graduate student. The student, Madhav Gadgil, had just finished his first semester at Harvard—this day was the end of exams. He had traveled thousands of kilometers to come to Harvard with his new wife, also a graduate student. Although he had always been a good student, he was worried. When he had decided to leave India for graduate study in ecology at Harvard, "everybody said 'don't apply, they won't admit you.' I thought they were unduly pessimistic."[1] Gadgil had been correct—Harvard admitted him. But now, would he do well enough to stay or would he have to return home, an overreaching Icarus who had flown too close to the intellectual sun of Harvard's ecology program?

The professor in the elevator was E. O. Wilson, "the brightest young star in the ecology-evolution end of biology at Harvard at that time," and the professor of Gadgil's first biology course at Harvard, evolutionary biology.[2] Wilson, as Gadgil explained, though polite to a fault, "was considered very remote, actually, by most people. He is a private person. He doesn't spend time with people." Surely the elevator must have felt small to Gadgil. Wilson was, after all, one of the most renowned ecologists in the world, even in his thirties—the man who had gotten tenure at Harvard more quickly than Watson, the codiscoverer of DNA. And Gadgil, as he had admitted, was not at all sure how well he was doing in his classes. As the doors closed and the elevator began to move, Wilson turned to Gadgil, smiled his charming Southern smile, and (as Gadgil wrote nearly thirty years later with the calmness of a sprinter remembering his first race) "He complimented me on my excellent performance" (26). The elevator, we may assume, was going up.

Gadgil claims "from that moment onwards, I have never again doubted myself" (26). He has certainly gone on to have a remarkable career, most notably, founding the Centre for Ecological Sciences (CES) in India, where Gadgil has worked to form a world-class ecological science program in India that is firmly connected to an international network of ecologists and ecological institutions. Like Sálim Ali at the BNHS decades earlier, Gadgil returned to India trained in the latest theories and techniques of ecology. Unlike Ali, however, Gadgil had completed his training for a Ph.D. and was prepared to formally train other young ecologists. All but one of the scientists on the faculty of the CES in 1998 had been students of Gadgil's. They are internationally published and recognized, and their work is connected with worldwide networks of funding and research. At the same time, CES scientists strive to create a version of ecology that is oriented to local Indian (rather than just U.S. or Western) concerns and problems.

To Harvard and Back

Madhav Gadgil, like many of India's early political and academic leaders, was born to privilege. Of his father, Gadgil writes, "He took money and social status entirely for granted" (26). More directly, Gadgil told me, that, because his family was relatively well-off, "my parents did not think that it was important to worry too much about getting into a career in which you would earn a good living." Gadgil claims to be of no caste: "my father was utterly free of caste, class and religious prejudices. He refused to perform the thread ceremony for his sons, so I can proudly claim to belong to no caste whatsoever" (27). The thread ceremony is a ritual of the Brahmin (upper) caste. While it is admirable that Gadgil's family chose not to participate in the caste system, this denial of caste is most feasible for someone like Gadgil who continues to be perceived as upper-caste, whether he wishes it or not, by those people who do still care.

Gadgil's father had attended Cambridge University and was one of India's most renowned economists. He was an institution builder and reformer who had worked closely with farmers' cooperatives. Gadgil grew up in an explicitly political environment and his later life reflected a commitment to public service and improving his homeland. As a professional, Gadgil has even delivered public lectures in Marathi, the state language of Maratha, in which Gadgil was raised.[3] A public intellectual, Gadgil's father valued the life of the mind, and encouraged Madhav's interests in a wide range of subjects, but "given my fascination for wilderness, biology was the natural choice" (26).

Gadgil grew up in Pune, an inland city located a couple of hours by train from Bombay, in the hills of the Western Ghats. Pune is by all accounts a beautiful town and was a center of the early Indian independence movement. Gadgil was an avid bird watcher and wanderer during his childhood, taking his family's dog on strolls through the hills. Unlike Ali and many of the other early ecologists throughout the world, however, he did not learn love for nature from behind a gun. Gadgil's father encouraged his interest in birds and even introduced his son to Sálim Ali. Gadgil "corresponded with Sálim Ali when I was still in high school and college," and admired Ali's work on the natural history of birds.

When Gadgil went to college he turned down a place in the prestigious Pune Medical College (in India students enter medical school directly, they do not first complete a separate undergraduate degree) in favor of pursuing a degree in biology, but "it took me three years to leave behind [the] constant social pressure" (26). Having earned a bachelor's and master's degree, Gadgil wanted to go abroad to get a Ph.D. in ecology. He was certainly encouraged in this by his father, who had himself studied at Cambridge. Gadgil also claims that there was a "major lacuna in my education, which was that the Indian educational system does not allow you to simultaneously study mathematics, statistics, and biology." By 1965, when Gadgil completed his master's degree and entered Harvard, he had seen publications from overseas showing the newest trend in ecological research, the use of mathematical modeling to explain complex natural systems. Thus, in stark contrast to Ali, who left college in 1914 because of his dislike of the small amount of math then required of biologists, Gadgil left India for his Ph.D. in 1965 because Indian ecology programs did not include enough math. The Indian educational system had not changed; what had changed was the science of ecology itself, as it became a more mathematically sophisticated science.[4]

Gadgil Goes to Harvard

When Gadgil arrived at Harvard, E. O. Wilson was completely unknown to him. He intended to study marine ecology with Giles Mead, curator of fishes at the Harvard Museum of Comparative Zoology. Gadgil had met Mead in Bombay when he passed through as part of an international oceanographic expedition working in the Indian Ocean. Mead had actively encouraged Gadgil to apply to Harvard, and Gadgil had listed Mead as his potential advisor on his application. But in his first semester at Harvard, Gadgil took Wilson's course in evolutionary biology. At this very time Wilson was developing his ideas on theoretical population biology, and he shared them with his graduate students. Gadgil was "enchanted" (26).[5] This was the course for which Wilson had complimented his "excellent performance." In contrast to this thrilling experience with one of the world's great young ecologists, Mead "was strictly a naturalist, knowledgeable about fishes from the deep seas; but not really of the exciting developments that were taking place in ecological theory" (26). After a brief summer trip on an oceanographic vessel during which he dutifully attempted to come up with a research plan, Gadgil jumped ship (as it were) and decided to concentrate on theory. He joined Wilson's group of young theoretical ecologists, working directly under Wilson's former student William Bossert, by that time also on the Harvard faculty.

Gadgil's dissertation, produced in four years, was, he asserts, the first thesis submitted by a biology student at Harvard to be based on mathematical modeling. Bossert, Gadgil's advisor, had done a modeling dissertation earlier, but he was not enrolled as a biologist—instead he was a crossover from the applied mathematics department. Gadgil's work on life-history strategies (in what activities a given animal should invest its time to get maximum growth and reproduction) met with widespread approval from ecologists at Harvard and elsewhere, and it continues to be cited by practicing ecologists thirty years later. Gadgil's project was at the very forefront of theoretical ecology when he finished it in 1969. It was a project that Gadgil claims "had been opened up by the development of modern computers" (27). Gadgil was among the first generation of ecologists to use computers for research.

There are several striking parallels between the early careers of Gadgil and Ali. Both men were spared financial worries by their family's wealth and status.[6] Both benefited from familial connections at an early age, with Ali taking his uncle's introductory letter to the BNHS offices and Gadgil being introduced to Ali by his father. Both were urbanites but lived on the edge of town and were in a physical and social position to learn early in life to view the forest as a place of recreation and leisure rather than of work or danger. Both came from the state of Maharashtra and from cities that had been heavily influenced by the British (Bombay was literally a creation of the British Empire, while Pune had long predated European rule but was used by the British to escape the coastal heat during the summer). Both found the educational offerings in ecology to be lacking in India, and so both looked overseas. And by nearly random chance they both ended up working with people who were at the center of the newest developments in ecology for their generation.

Of course, randomness can be overstated. Gadgil, even if he did not know E. O. Wilson, did know Harvard. There was not much chance of Gadgil applying to a second-tier university, nor was there much chance that Wilson would be based there. And global politics and economics certainly influence where leading universities emerge across the world and which ones can attract and afford to support international students. Still, there were several universities with excellent reputations; for example, Gadgil could have chosen his father's alma mater. He would not have had anything like the training he got at Harvard.

Harvard Ecology in the 1960s

Harvard was a special place for ecology in the latter half of the sixties. As with Stresemann's lab in Germany in the twenties, Wilson and his colleagues at Harvard had attracted a particularly motivated and bright bunch of graduate students who would proceed to make several important contributions to the ecological literature.[7] Wilson himself was in the midst of developing two of his most famous ideas at this time, sociobiology and island biogeography. Gadgil recalls that "before the theory of island biogeography monograph was published, Wilson gave a seminar. . . . I don't know how influential it was [to my own research] but I followed it with great interest." And perhaps most important for Gadgil, Harvard was also one of a handful of places where mathematical, or theoretical, ecology was being actively encouraged (another leader in this area was Princeton's Robert MacArthur, Wilson's coauthor for the island biogeography work).

The first big shift in ecology theory and practice had been the move from taxonomy to the ecology idea based on studies of behavior and interactions (in the natural history tradition). The second was the shift, led by scientists such as Ernst Mayr at Harvard (Ali's former colleague in Stresemann's lab), from this observationally based ecology to a more theoretically informed evolutionary ecology. This move combined behavior, genetics, and taxonomy into one enterprise—the attempt to understand the variety of life using the insights gained from evolutionary theory. All animal behaviors, in the adaptationist model, can be understood and explained not just as idiosyncrasies but as adaptations that provide some benefit or cost to the organism that is thus naturally "selected for or against" versus those organisms that do not practice the behavior.[8]

For instance, small fish that swim in schools live longer than fish of the same species that do not. An evolutionary biologist might argue that schools of small fish have the chance of looking big to a predator, that there are more eyes to watch for predators, and that if one does attack, each individual's chances of getting eaten are relatively smaller. Over time, then, the solitary fish are eaten and can't reproduce, while schooling fish survive and pass on this behavior to all their offspring (assuming it has some genetic basis and is thus heritable). Eventually, almost all the small fish will swim in schools. Evolution is a dynamic process, though, and at some tip-over point, there will be so few solitary small fish that it is a bad investment for predators to hunt them, and the solitary fish will be able to hide quite successfully, as opposed to large, flashy schools of fish. Thus, there will always be some low level of small independent fish. The process is dynamic, with populations of schooling versus nonschooling fish in flux over time. In this fashion, when evolutionary ecologists look at a biota, they assume that all the interactions within that community are the result of evolved changes over time.

Evolutionary ecology was a powerful new tool for explaining plant and animal characteristics and behaviors. Ecologists began to look not just for how organisms interacted, but also for the potential ordering logics behind those interactions. It was here that theoretical (mathematical) ecology entered. With the aid of the computer revolution, which was just beginning in the sixties, ecologists could design mathematical models of ecological systems to predict patterns of change or stability. Such modeling allowed them to predict what would happen if very small behavioral change occurred over the course of thousands of generations, and ecologists could for the first time conduct computerized "experiments" that simulated evolutionary (geological) time.

This was the type of work Gadgil did for his Ph.D. As he had hoped in India, he was able to devote almost all his class work at Harvard to mathematics. Even with this coursework, Gadgil was attempting something with his dissertation at the very limits of his mathematical abilities, and which most biologists at the time were simply not mathematically sophisticated enough to imagine. He claims, "I had the courage to [pursue that dissertation topic] largely because in my wife Sulochana [also enrolled at Harvard], I had a first-rate mathematician I could consult." (27). Using mathematical modeling, Gadgil proposed a model for life history strategies (when an organism devotes its energy to growth, reproduction, etc.) based on trade-offs and the optimization of resources. The individual organism may not be aware of how to optimize resources, but over time evolution will "reward" (allow to reproduce successfully, many times) those organisms that (randomly) optimize their resources; similarly, evolution will "punish" (kill and not allow to reproduce successfully) those that do not.

In 1969, when Gadgil finished his Ph.D., he was among the most highly trained young ecologists in the world.[9] At that time, Wilson offered him an assistant professorship at Harvard, something he had occasionally done with his very best students. But in case Gadgil wanted to leave the town of Cambridge, Wilson's colleague and friend Robert MacArthur offered Gadgil a similar position at Princeton. Gadgil turned down MacArthur, and spoke with Wilson, saying he was reluctant to take an assistant professorship, as he and his wife would likely leave in one or two years. Wilson then offered a lectureship, "Because then in some sense you are not morally committed to stay for any length of time.'" Gadgil accepted this offer.

As Gadgil admits now, "my wife had an offer at MIT, and I had one not only at Harvard but at Princeton also, so they thought that we were mad going back to India without a job. But my wife was even more than me convinced that she wanted to come [back to India]." So in 1971, after six years in the United States and two years as a lecturer at Harvard, Gadgil returned home to India: "There was a farewell party, and Wilson joined the party, which was in the evening. People who had been at Harvard for years said . . . 'He must really like you . . . he has never come to any party of an evening.'" The first time I met E. O. Wilson in person, at Cambridge in the late spring of 1999, I found Madhav Gadgil sitting with him, drinking tea. They were saying fond farewells as I arrived, Gadgil loaded down with a bag of new books from Wilson, Wilson proudly looking at a stone tablet of at least forty pounds carved with a traditional snake motif brought from India by Gadgil. Wilson turned to me, asking how I had first met "the illustrious Madhav Gadgil."

Harvard Ecology in Bangalore: The CES

In late May 1998, just before the monsoon arrived, the trees of Bangalore exploded with blossoms—fiery reds and oranges, swirling yellows, pinks, and whites, vivid purples. Their fragrances filled the air, overriding the smell of auto exhaust. (In much of southern India Bangalore is known as the Garden City.) While the rest of southern India sweltered in the hottest part of their year (heat-related deaths were reported each day), Bangalore dwellers (at a thousand meters above sea level) had the luxury of complaining about daytime temperatures approaching ninety degrees. This agreeable city is hardly a hidden gem—the city's tourism office claims Bangalore is the fastest growing city in Asia (six million people in 2003). The spacious streets of "India's yuppie heaven"[10] are filled with shining new auto-rickshaws and, unusual in India, women in blue jeans driving scooters. Bangalore, also called India's Silicon Valley, is a preferred site for multinational businesses. Finally, it is the home of India's most prestigious scientific institution, the Indian Institute of Science.

Surrounded by a high wall, the huge campus of the Indian Institute of Science (IISc) is almost a small town in itself, with a security force, a small store, several cafeterias and teahouses, a travel agency, bookstores, a gymnasium and recreational center, a few small forest tracts, and the homes for almost the entire faculty and many of the staff. At dawn hundreds of people jog or walk along its many roads, and many lab lights burn late into the night. The campus crackles with intellectual energy. The teahouses are filled with animated discussions, as is the Indian norm, but here the topics range far from cricket or football (though they too are discussed). Set in a simple building five minutes walk from the center of the campus is the Centre for Ecological Sciences (CES). It is here that Madhav Gadgil can now be found.

When the Gadgils returned to India in 1971, they knew they would be able to at least get a living wage from the government while looking for permanent jobs. The government had set up a program under the auspices of the Council of Scientific and Industrial Research to support Indian scientists who had Ph.D.s but who were temporarily without a job. The CSIR offered them a type of automatic postdoctoral fellowship. The Gadgils elected to stay and work in Pune (their home) while looking for more permanent employment. During this period in Pune, Gadgil worked with V. D. Vartok on sacred groves in the Western Ghats—an experience that would shape the way that he developed his research interests particularly in the latter half of the eighties and the nineties (see chapter 8).

The Gadgils' break came two years later, in 1973, when Sulochana was recruited for the IISc's Centre for Theoretical Science (CTS), whose goal was to bring together "people with an interest in many different scientific disciplines, but a common interest in mathematical modeling." Not only was Sulochana, with her background in math and its application to meteorology, a good fit for the CTS, she also persuaded the CTS to offer her husband a position.

Madhav Gadgil's work at Harvard using math and computer modeling to help understand the dynamics of the natural world was exactly the sort of biology that interested the director of the CTS, which was attempting to use mathematics as a unifying theoretical base for all sciences. There was before then no other ecologist on staff at the IISc—it was not considered a "hard" science. There were molecular biologists, but no organismal biologists. Mathematical ecology was not the sum total of Gadgil's interests, though. He was also interested in field ecology (as with his sacred grove studies around Pune), and he feared becoming a "desktop ecologist" who only worked with computers and simulations. Further, he wanted to work with other ecologists, not just with mathematically interested colleagues in other disciplines. Gadgil said that "right from the beginning [the director of the IISc] helped to try to get the infrastructure built up and get the funds to start a college [of ecology] here." Within five years, by 1978, proposals had been sent to the government's Department of Science and Technology for the formation of the CES. And in 1983 the CES was officially constituted as a part of the IISc and Gadgil was named its chairman.[11]

Gadgil had not waited for the formation of the CES to begin reaching out and training potential colleagues. A steady stream of visitors, primarily professionals interested in wildlife biology, came to the CTS to talk with Gadgil about research design and ecological field techniques. Gadgil's arrival at the CTS in 1973 coincided almost exactly with the government's decision to freeze the PL 480 funds, thereby restricting the flow of U.S. "experts" into India. Because the government of India was making it increasingly difficult for outside ecologists to work there, Indian ecologists and field biologists, who had few other avenues for information, sought out Gadgil, aware of his familiarity with the newest ecological theories and techniques. Gadgil had also begun recruiting graduate students to work with him—some came to the IISc to study with Gadgil, some others had come intending to do something else before Gadgil turned their attention.

Raghavendra Gadagkar was one of the latter: Gadagkar began his Ph.D. work on molecular biology in 1974, one year after Gadgil had joined the CTS. During his first year, Gadagkar read a paper that Gadgil had published on social wasps and realized that Gadgil was also at the IISc.

> I went in and met him and, in fact, initially I said, "I know of wild popu-
> lations of these wasp colonies. If you are interested in studying them, I can
> show them to you." He said, "No, I don't have time to study them, but if
> you are interested in studying them, I can help you!" . . . He inspired me
> to look at the wasps more scientifically. And he opened my eyes to all the
> literature at that time. Because he was a student of E. O. Wilson at Har-
> vard, he was familiar with this.[12] So even the five years when I was a
> Ph.D. student [in molecular biology], from '74 to '79, I really spent quite a
> bit of time, a criminal amount of time, in the CTS. . . . I read Wilson and
> Hamilton and Dawkins and Trivers and all these people and became very
> interested. . . . I was doing serious work collecting data and reading these
> papers, and attending meetings. . . . I used to tell my Ph.D. supervisor that
> "it is just a minor hobby."[13]

When Gadagkar completed his Ph.D. (still in molecular biology) in 1979, Gadgil arranged for the CTS to offer him a postdoctoral fellowship so he could write up and publish his wasp behavior work. Instead of just writ-ing, though, Gadagkar kept making more observations of the social be-havior of the wasps, and his fellowship kept being extended until he joined the CES as a faculty member in 1983. His hobby had become a full-time job.

A second staff member at the CES, N. V. Joshi, was Gadagkar's room-mate in graduate school. Joshi received his Ph.D. in molecular biophysics, studying the optimum shapes for molecules: "Whatever biology I learned, or ecology, was by talking to Professor Gadagkar. And he and Professor Gadgil were doing some work together on wasps. So that is how I came to know the CTS group. So it was mainly because of my being a room-mate of Professor Gadagkar, that was, how should I put it, 100% responsi-ble for my coming to ecology." In 1978, while Joshi was completing his Ph.D., a position opened up at CTS for a programmer. As Joshi had been working extensively with computers for his research, he was accepted for this job. "It so happened that most of the people who came there were ecologists. They were kind enough to interact with me not as a program-mer, but as a colleague. So to that extent I started doing statistical analysis and simulations, things of that kind, oriented towards ecology."[14] After Joshi finished his Ph.D., he never again did molecular biophysics. He too had been converted. In 1978 the CTS was almost certainly the only place in India, and one of the few places in the world, to have a full-time computer modeler and statistician working with ecologists on quantitative ecology.

The CES is a small institution, with five full-time faculty members when I visited in 1998 (one additional member has since joined). It is somewhat surprising that two of them received their Ph.D. in subjects other than ecology. This reflects Gadgil's emphasis on building a quantitative theoretical ecology program at a time when most ecologists (not just in India) were quite uncomfortable with mathematics. Scientists trained within other scientific disciplines were much more prepared to incorporate computer simulations and mathematical modeling into their work than traditional ecologists or organismal biologists. As he had told the director of the CTS when he was first hired, though, Gadgil was interested in both theoretical ecology and in field ecology informed by theoretical ecology. He was committed to the idea that even good field biology could benefit from quantification and statistical analysis (when appropriate). While Joshi is purely a computer ecologist, and Gadagkar's research is now on wasp colonies in a screened enclosure on the roof of the CES building, the other two faculty members at the CES practice Gadgil's vision of theoretically informed field ecology: Raman Sukumar was one of Gadgil's first Ph.D. students (studying human-elephant interactions); Renee Borges (studying habitat fragmentation, also the first female faculty member) received her Ph.D. at the University of Miami in 1989.

The faculty members at the CES have a strong sense of their institution's uniqueness. Sukumar's comment is typical: "Professor Gadgil was the first [in India] to really . . . set up an ecology program based on more modern ecological theory, and [give] an evolutionary biology approach to ecology." Gadagkar echoed this: "Madhav Gadgil has been very instrumental in bringing modern ecology, quantitative ecology, and animal behavioral ecology, in making it happen here [India]." When pressed to define *modern ecology,* Sukumar turned to specific theories: "A lot of theories in community ecology, for instance, island biogeography, even population dynamics, theories of life history strategies, R and K selection—this Professor Gadgil had worked on earlier. . . . Of course, the whole evolutionary biology approach to it." The theories that Sukumar selected as examples of modern ecology were all being developed and worked on at Harvard when Gadgil was there in the late sixties and early seventies.

The dedication among the staff of the CES to their self-image as modern ecologists reflects their determination to participate in the international practice of ecology, a practice dominated by the paradigms and theories of U.S. ecologists drawing upon the evolutionary ecology synthesis. Without a doubt the CES succeeds in this. All the faculty have published in prestigious Western journals. And though the CES is fairly small, it pulls in a large amount of money from organizations like the MacArthur Foundation, the Ford Foundation, and the Smithsonian Institution. The faculty travels frequently in the West. Admissions at the CES are highly selective; there were eight graduate students in 1998. Gadagkar's last book was published by Harvard University Press, and Sukumar's by Oxford. Within India, the scientists of the CES are viewed by other ecologists and wildlife biologists as being highly theoretical, oriented to pure research, quantitatively savvy, and familiar with the most complex of current ecological ideas. It is striking how often, when questioned about their research projects, ecologists or conservation biologists in places outside Bangalore are self-deprecating about their research and training—there was a sense that "we get by—but at the CES, they do the complex quantitative ecology that you [in my research] are interested in."[15] Like Sálim Ali and the BNHS in the fifties and sixties, the faculty of the CES define themselves as carriers of the new into India—and the new for the CES is the brand of ecology developed by E. O. Wilson at Harvard.

Imperial Harvard Biology or a Harvard-India Hybrid?

It is possible to argue that the science of ecology as practiced by the scientists of the CES is an import devised in the United States and culturally inappropriate for India. One of the basic insights derived from the social studies of science in the last thirty years is the recognition that science gets practiced in specific historical and cultural milieus. This is not a radical suggestion, though furious debates between scientists and the academics that study science suggest that the matter is controversial.[16] To say that science is localized does not mean that a Kenyan scientist timing a certain bacteria's replication will get a different result than a German scientist would, or that a Canadian astronomer using the same equipment would see a different set of Jupiter's moons than Galileo. Rather, what is meant is that the questions that scientists ask, the methods they use to answer those questions, how they interpret their results, how they get funding for their research, and who listens to (or, more relevant, who publishes and reads) their research are all determined in large part by the historical and cultural context in which the scientist works and lives. This, of course, is not true of science alone but of all scholarly inquiry.

It is easier to see the import of this point with an allegory. Imagine a wealthy society that has traditionally believed that moths are physical embodiments of the sacred. Perhaps the priests in this society even assert that moths are the souls of ancestors. Nowadays, of course, most of the people of that society—let's call it Mothopolis—realize that moths are really just insects, albeit extraordinarily beautiful insects. Still, moths exert a tremendous pull on the collective imagination of that society. On Mothday, celebrated annually for hundreds of years, everybody hangs fresh wool banners in their back yards upon which moths can feast for the coming year. The printed money of Mothopolis is adorned with moths of many shapes and colors. Little children pretend to be moths. On roads not equipped with enclosing moth-proof screens, speed limits are set low enough so that moths are not injured by moving cars. (Of course, Flit and blue-light insect zappers are not sold anywhere.) Now imagine that scientists discover a previously unknown mite that parasitizes moths. During the national news, there are moth mite stories. Across the land, scientists take up the study of moth mites, studies that are well funded by both private organizations such as Moths Unlimited (MU) and the Fly Free Forever Fund (FFFF), and even by the government's national science board. When complete, the studies are published in scientific journals and reviewed in the popular press as well. And when it is reported that one of the scientists had chosen to conduct his study by euthanizing thousands of moths, as opposed to merely temporarily stunning them to collect the mites, there are massive protests by MOTHS (Moths Ought To be Handled Sympathetically) and the government takes steps to ensure certain standards of research accountability.

In the real world, our world, there are mites that live in moth ears. This is not something that is widely studied, however. As a society, we don't care very much about mites in moth ears. And because we don't care, such research is difficult to fund, is published in specialized journals not read by the general public (or even most scientists), and is seldom even conducted. There is no reason not to study moth mites, as opposed to bald eagles or white-tailed deer, other than what scientists get interested in, and what they become interested in is a product of their cultural-historical context.

Now, imagine that the wealthy moth-loving society, Mothopolis, learns that in a less wealthy neighboring country, Lotsofmoths, there are thousands of beautiful new moth species. Scientists from Mothopolis stream into Lotsofmoths, establishing moth research sites throughout the land. To the horror of the Mothopolitans, many people in Lotsofmoths are indifferent to moths and even exterminate them on sight or use mothballs to keep them away. Of course, some people in Lotsofmoths have always liked moths—after all, there are lots of them around, and if the Lotsofmothians had wanted to they could have eliminated all the moths. Research money is given to Mothopolitan scientists to do moth research in Lotsofmoths, and even more money is given to the government of Lotsofmoths to pass laws conserving moths and to begin covering the nation's highways with moth screens. The most talented of the moth-loving young students in Lotsofmoths are encouraged to go to universities in Mothopolis, where they are given scholarships to be trained in moth science and conservation. They do this and then return to their homes in Lotsofmoths to study moths and to publish their results in the prestigious Mothopolitan journals.

Likewise some scholars in our world have moved from recognizing that science is culturally situated to claiming that a science grounded in one culture can be imposed on another culture when great differentials of wealth and power exist.[17] Ecology as it is practiced at the CES can be seen in this light as part of a process whereby U.S. (scientific) ways of understanding the natural world, bankrolled by U.S. wealth, have been imposed on India.[18] This interpretation, and my moth story, however, are far too simple. A more subtle account would be one in which traditional Lotsofmothian ways of relating to nature are challenged and modified by new—but not completely unrelated—ideas developed in a different cultural context.

This is the story that is told by some historians about smallpox control in India. Historians have shown that at least as early as A.D. 1700 Indians had traditional forms of smallpox control, including a type of immunization that was centered on religious beliefs. These traditional beliefs were overturned by British physicians, who appropriated the immunization technique and then developed a form of smallpox control based on Western notions of disease and contagion. The British did not introduce smallpox control to India (though many British thought they did so), rather they introduced a different way of conceptualizing that control and, through experimentation, they modified immunization in ways that made it more effective.[19]

Still better stories (i.e., more complete and nuanced) about the relationship between science and scientists in India and those in the United States can be told by combining historically rich descriptions with the words of the scientists themselves. For my research, this double method can be realized literally, in the case of still living people who allow themselves to be interviewed, and through the careful study of how scientists pursue their research goals and practices. Documenting institutional and financial ties and relationships is still a crucial part of the context in which research occurs, and that is why I have devoted the first half of this chapter to discussing the links between the CES and Harvard. But there is more going on than "cultural imperialism"—the imposition of Harvard concerns onto southern India. Gadagkar and Sukumar, both faculty members at the CES and both Gadgil's students (one informally, the other formally), practice an innovative type of Indian ecology on the border between locally generated concerns, and questions generated by global research agendas.[20] They are neither resistance fighters against the colonizing West nor are they imperial agents. They are also not value-free adjudicators of scientific results. Instead, I see them as scientists located at a complex site that is simultaneously peripheral to First World concentrations of ecological power and wealth yet is also producing globally respected ecological work. At present they are leading other ecologists in new ways of doing and conceptualizing ecological work.

Raghavendra Gadagkar

Raghavendra Gadagkar, the molecular biologist who turned his hobby into his vocation, is the world's expert on the social wasp *Ropalidia marginata*. Gadagkar has used this primitive wasp species to study the origins of social behavior in insects: he considers himself a behavioral ecologist. He has a contract with Harvard University Press to prepare a monograph summarizing his career of research on this one species. His research has been published in the leading journals of behavioral ecology in the United States, Europe, and India over the last twenty years.[21] He regularly attends conferences in Europe and the United States, and his recent summary of the field of animal behavior (also published by Harvard) was hailed by E. O. Wilson as "a highly readable update of the spectacular evolutionary productions of animal social behavior. The author . . . ranges smoothly from the natural history to the genetic basis of the many phenomena that have surfaced during the past two decades."[22]

The wasp that Gadagkar studies lives in colonies. All the females in a colony are capable of reproducing, yet only the queen actually does so. One of the central questions for Gadagkar's research has been to discover why nonqueens, capable of reproduction, would stay in a colony to help maintain offspring which are not theirs—are these female wasps altruistic? The usual explanation is that all females in a given colony are closely related. By furthering the success of a close relative's offspring, the female wasp who does not reproduce still is furthering her own genes, and doing so more successfully than if striking out on her own. When Gadagkar did genetic tests on colonies of wasps, however, he found that the wasps in a given colony were not as closely related as had been expected. This suggests that there must be some other evolutionary explanation for why wasps develop "altruistic" behavior. The search for this explanation drives the research of Gadagkar and his students, research that has ranged far into other biological disciplines.[23]

Gadagkar's work on the social behavior of *Ropalidia marginata* requires him and his graduate students to go no further than a rooftop enclosure at the CES that houses a series of containers for his many wasps. He needs no equipment other than his computer to run the statistics and modeling programs and a pencil and paper to record his (or, more often, his students') observations. This, Gadagkar told me, is the perfect kind of science for India. It requires creativity, intelligence, and an awareness of the relevant literature, but it does not require expensive equipment or huge laboratories. Gadagkar has made the same argument in print: "ethology is just the right choice for young biologists embarking on a research career in India, because of the easy access we have to an incredibly rich fauna and flora to use as model systems. [As it] seldom requires very expensive and/or imported equipment or chemicals . . . ethology provides realistic opportunities for Indian biologists to provide international leadership" (10). Gadagkar himself is the proof of this contention. He works on a wasp species indigenous to India, a species whose very accessibility is what got him started on the project in the first place (in 1974, when he offered to show Gadgil a large colony of wasps on the Bangalore University campus). His project was so inexpensive that he could conduct the research without a budget while technically a graduate student in another department. Through continued hard work he has become an international leader in the field.

Gadagkar is vocal about the difficulties of doing what he calls "internationally competitive research" in India. The most obvious of these difficulties are "Indian conditions of technology and economy," difficulties that are particularly acute in fields such as molecular biology. Gadagkar claims that Indian scientists are obviously capable of doing excellent research, but they "find themselves in an economically and technologically backward environment that does not give them a fair chance to compete with their [Western] colleagues."[24] Lack of funding and equipment become problems, though, only because Indian scientists insist on working on scientific questions similar to those worked on in the West. Gadagkar claims that in the conduct of basic science, "it does not matter that much which field we work in," as long as "you work at the cutting edge of human knowledge" (45). The true problem, according to Gadagkar, is the tendency of Indian scientists to work on topics judged as trendy by Western scientists rather than on problems that could be easily studied in an Indian context.

Almost invariably, Gadagkar associates trendiness in biology with molecular biology—this is why he can feel that his research on insect behavior, so strongly based on Wilson's continuing research on ants at Harvard, is not based on a trendy problem. He complains that, for instance, "most of our biologists prefer (it is not so much a preference as it is inertia) to work on a bacterial or virus strain brought from the U.S.," as opposed to working on India's "incredibly rich fauna and flora" (46). He also spoke of the danger of scientists "jumping on the bandwagon" and abandoning that which they could do well:

> There are little colleges in India where there is considerable expertise in doing some kind of research. Let us say histology. You . . . have this thing called a microtome, so you cut sections [of a specimen] . . . and you describe it [using a microscope] and take pictures. It is as much an art as it is science. And people sometimes have spent their whole lives doing that, and they are really experts. . . . [B]ut histology is not considered fashionable anymore. So they throw . . . away their microtomes, throw away histology, and they try to do, for instance, molecular biology. And the level of molecular biology that they are able to do is far below the histology, because it is so expensive and it is a new technique they have to learn. So, from being first-rate histologists, they become second-rate molecular biologists. And now, for example, if I, as part of my animal behavior work, find that an animal is using a certain gland to communicate with another animal and I want someone to take a section of that gland and describe it to me, one would have thought that there would be any number of people in this country who would have been able to do it for me. But they have all thrown away their microtomes and are trying to sequence DNA, and not doing it very well either.

This means that these scientists are always one step behind, taking up today what was current in the U.S. or Europe yesterday. According to

Gadagkar, this failure to work on manageable problems with manageable technology "allows Western scientists to set agendas. Indian scientists then inevitably end up following the West's lead and seldom provide the lead themselves" (45).[25] Gadagkar does not use the language of imperialism, but historians and social scientists would see in Gadagkar's statements a plea for Indian scientists to decolonize their minds.

Intriguingly, Gadagkar sees his type of ecology as constituting a liberation from Western agendas. Although his research is extraordinarily well received in the United States and Europe and his topic, behavior in social insects, has been popular since Wilson's pathbreaking work in the seventies, Gadagkar's specific research project has focused on a previously little studied wasp species and represents an original contribution to the animal behavior literature. Gadagkar sees his success as an example of how Indian scientists can choose research topics well adapted to India—ones not being actively considered in the West—and through good work move the scientific debate to include this new, Indian, angle. Gadagkar admits that physical isolation is sometimes a problem, in that there is literally no one else in India who is doing similar work and with whom he can have a technical conversation, but he claims that the new information technologies of e-mail and the Internet are removing these barriers. A more serious barrier to Indian work on the international periphery is anonymity: the lack of recognition of one's institute or name when submitting articles for publication and the failure of foreign colleagues to cite one's work because they are unaware of it. These problems, Gadagkar claims, diminish somewhat with seniority, but they make it difficult for a new scientist in India to break into the Western scientific conversation.

Gadagkar thinks the practice of science, the scientific way of knowing the world, can be applied in any culture. It is particular research programs and research interests that are culturally specific (it is not science that is Western—it is choosing to study moths, for example). He said, "I think that the way I do science is certainly influenced by the fact that I was born in India. I live in India; I am subtly influenced by that. I do not believe that science is completely international and beyond all values and nationalism." Although he thinks that cultural influences on science are weak, largely limited to the economic and technological development level of a society, at this mechanical level he does admit that context matters. Further, as seems obvious from his frequent use of terms like *backward* to describe Indian conditions, he is not a critic of Indian development and modernization. And though Gadagkar wants Indian scientists to not orient themselves to Western problems and trends, he escapes working on a trendy Western topic only by virtue of his study species, not his overarching questions.

Gadagkar is interested in another aspect of ecology beyond animal behavior, the study of insect diversity. Because of his work on the social behavior of wasps, he has developed a wider interest in other insects. He was soon "struck by the fact that nobody in India worked on insect species diversity, and they are supposed to be the most diverse group in the world, and the tropics, including India, have a very large number of species!" When Gadagkar asked entomologists why this was so, he was told that Indian scientists did not have the necessary resources to do thorough insect studies: "this is not something we [Indian scientists] can do very easily, because very different kinds of research is done in other stations, the Smithsonian and all. For example, people set up very powerful light traps in the middle of the forests, and those light traps run for twenty years, and they have a huge army of people to collect all the insects, and there is another army of staff who sort them out, and so on. It can only be done by these very large institutes." It is true that many of the insect species studies conducted by the Smithsonian in Central and South America have been massive undertakings. Terry Erwin became renowned among ecologists in the eighties for his tented trees, surrounding entire rain forest trees with fabric and gassing them to collect all the insects on the tree.[26]

Gadagkar admitted that some of these techniques were not feasible for India, but he does not think that such big-budget techniques are necessary to study insect diversity: "I decided to remove this stumbling block—this perceived stumbling block—so I set out to develop a package of methods for studying insect diversity which did not need any of these things—was low cost, indigenous, inexpensive, a package which could easily be applied and relied more heavily upon man power than on technology." He devised several cheap alternatives to the Western technologies. These included battery-powered seven-hour light traps that could last through most of one night (but would have to be changed each day), pitfall traps made of simple plastic containers placed in the ground, and containers of detergent and water hung from trees. These methods all relied on frequent monitoring, whereas many of the Western techniques could be set up in a rain forest and left unattended for months. Gadagkar's insect traps had to be changed daily in most cases, but labor costs in India are low. In the CES annual report for 1990–91, a document essentially prepared for the Ministry of Environment and Forests, whose funds support approximately 50 percent of the CES projects, he claimed, "We believe that the methods devised by us will enable Indian ecologists to undertake interesting work in the area of insect species diversity patterns."[27] He also wrote that "the number of biologists is negatively correlated to the number of biological species in different regions of the globe. . . . [T]his may be partly attributed to the relative economic backwardness of tropical countries, the lack of facilities for research and sometimes to the lack of the tradition of modern scientific work. We [the CES] believe, however, that at least sometimes this is due to the lack of appropriate research methodology suitable for tropical conditions" (28). Indeed, one of the CES graduate students, T. R. Shankar Raman, had amassed an impressive and quickly growing insect collection from traps scattered deep in the rain forest of the Kalakad Mundanthurai Tiger Reserve.

By developing these techniques Gadagkar hoped to encourage work on insect diversity. On a larger scale, he believes that taxonomy is another field where Indian scientists could make a significant contribution to international science. Gadagkar has written a number of short popular articles for student-oriented magazines in which he encourages students to undertake taxonomic research. He claims he is trying to "create public opinion, particularly among students. . . . I try to open the eyes of the other students to the kinds of problems that exist out there." In an article for the IISc campus magazine, Gadagkar asks whether biology should go "back to stamp collection" (a label applied by biologists, particularly ecologists, who strove to be more experimental and less oriented to natural history), or if we should "spend the rest of our professional lives staring at that interesting one kilobase DNA sequence isolated from one bug?"[28] Early ecologists in India, such as Sálim Ali, struggled to move the practice of ecology by Indians beyond collections and toward a more theoretically informed and experimental science. One reason Ali's collaboration with Ripley back in the fifties and sixties had been so important was that Ali was not assigned the taxonomic work—that had been Ripley's responsibility.

In his popular-press book *Biophilia*, E. O. Wilson suggests that developing nations should focus their ecological research on taxonomy: "Can there be an Ecuadorian biology, a Kenyan biology? Yes, if the focus is on the uniqueness of indigenous life. Will such efforts be important to international science? Yes, because evolutionary biology is a discipline of special cases woven into global patterns, nothing makes sense except in light of the histories of local faunas and floras."[29] Gadagkar has clearly been influenced by Wilson's writings: his article on taxonomy includes a long quote from Wilson decrying the extinction of biodiversity. Gadagkar also wrote a highly laudatory review of Wilson's autobiography for a student magazine, suggesting that Wilson should be a model scientist for young biologists. Intellectually, of course, Gadagkar's research is based on the insect behavioral ecology developed by Wilson. However, Wilson's suggestion that developing countries should focus on taxonomy has met with a fair amount of criticism.[30] Why should a Kenyan biologist be limited to working on Kenyan taxonomy, when Wilson can travel the globe looking for suitable ecological systems to sustain his theory?

Gadagkar explained that "this remark should not have been made by Wilson, it should have been made by us. Not just made, but I think that we should just do it . . . and show the world that we have done it. You see, if E. O. Wilson says that Indian scientists should do taxonomy, now of course, someone will say that you are preventing them from doing the sort of high science that is done elsewhere. So it should not come from there, it should come from us. I think that we must recognize where we have the advantages and where we have the disadvantages." Gadagkar recognizes that for Wilson to tell Indian ecologists what to study appears to establish a neocolonial relationship—a matter of violating political correctness to Gadagkar—but he has no problems with the content of Wilson's statement. The problem with taxonomic studies, though, is that they are always something that would be good for someone else to do. Ecologists in the United States are increasingly concerned because one effect of the low prestige of taxonomy is that millions of species (primarily insects) will become extinct before they are ever recorded. And if a species is not recorded it is difficult to save it. Gadagkar shares this concern: "Large scale destruction of habitat, especially forested habitats in tropical countries such as Brazil or India, are driving large numbers of species extinct—and extinction is forever."[31] Neither Gadagkar nor the U.S. ecologists, like Wilson, are willing to abandon theoretical ecology to take up taxonomy, though. Gadagkar encourages students to do so, and Wilson encourages the ecologists of the developing world to do the same. Neither will do it himself.

Raman Sukumar

Raman Sukumar was the third of Gadgil's graduate students, enrolling in the microbiology and cell biology department in 1979, before there was a CES, but working directly with Gadgil on ecology. Sukumar had come to work with Gadgil at the first "modern ecology" program in India, yet his dissertation research looked nothing like a U.S. research project. During his first year of course work he met with Gadgil to discuss possible research topics.

> [Gadgil] was talking about sexual selection in the peacock, the evolution of the peacock's tail, or social organization in babblers (a type of bird). There had been a lot of interest in cooperative living in the babblers. Another thing he suggested was that the BNHS had set up a field station at Bharatpur, and he said that there might be something interesting with birds there. And then I told him that my interests might be more with mammals. He said that the elephant-human conflict had not been looked at by anybody else. In Africa, a lot of research had been done on African elephants, but most of it had been on foraging and social behaviors, nothing really on elephant-human interactions.[32]

Sukumar took this suggestion and ran with it, producing groundbreaking research that suggested biological reasons for the specific types of problems associated with elephant-human conflicts in India.

Sukumar put the interplay between humans and elephants at the very center of his dissertation. Such research would never have been done in a U.S. ecology program. Ecologists there go to a lot of effort to find field sites that are as free of human disturbance as possible. And if an ecologist wanted to study a particular species, that species would be studied for insights into its "natural" behavior. Goodall studied chimpanzees in the jungle, not at garbage dumps where they forage. Schaller studied the prey species of tigers in Kanha National Park. Although he did look at deer in a zoo in Calcutta, that was only briefly and did not figure prominently in his findings.

Much of Sukumar's work looked very similar to the types of species studies that had been previously conducted. Sukumar attempted to get an idea of home ranges of elephants—"I had to be satisfied with crude estimates; I did not have facilities such as radio telemetry during my study."[33] He attempted to determine what elephants ate, their social organization, growth rates, population dynamics, and their migratory routes. All this is standard, solid animal ecology. Sukumar then took this information and tied it to human-centered questions: "Why is it that elephants come and raid crops, for instance? . . . [W]hy do elephants kill people, and what really happens in terms of habitats that are transformed by people? . . . [W]hat happens when people kill elephants?"

Sukumar was analyzing animal (elephant) behaviors and the ways these behaviors influenced how elephants responded to human interactions and modifications to the natural world.

His findings were fascinating.[34] Among his many concerns was an analysis of crop raiding in elephants. Crop raiding is a problem in India not just because of the loss of crops and income (which can be truly catastrophic for a small farmer—an elephant can eat a lot in a night) but because elephants kill humans in the ensuing confrontations. Sukumar's research was the first ecological analysis of the problem.[35] His results indicated that most crop raiding was conducted by a few solitary male elephants. When a male elephant enters musth, a hormonally controlled breeding cycle, the quantity and quality of food he eats determines how long he stays in musth. Evolutionarily, it is to the elephant's advantage to stay in musth as long as possible, so that he can continue to breed with female elephants. By raiding rice, sugarcane, maize, or wheat fields, an elephant receives excellent nutritional value for a minimum of effort. This good nutrition will keep the male elephant in musth longer.[36] This advantage to the male elephant offsets the risk of being scared off by human guardians of the fields or even shot. Males in musth are much more aggressive than those not in musth and are thus much more likely to attack humans protecting their fields. Female elephants, however, have no such adaptive rationale for taking the risk of crop raiding. Further, females usually move about in clans of mixed ages and are less likely to take actions that would endanger the group.

Ecologists in India and the West immediately recognized Sukumar's work as significant, although he "was not really sure how it would be received outside India, because I knew that there had been a lot of very detailed work on the ecology/behavior of just a single species [by] a lot of people from the West, and this was something which was very different in one way." While Sukumar had taken a risk in his focus on elephant-human conflicts, he was still studying a very charismatic species. Elephants are big moths in Mothopolis. This choice of a charismatic species might have helped him in garnering initial attention for an unusual project. His reception might have differed had he worked on interactions between humans and cobras or even rats.

In any case, Sukumar's work was well reviewed, and as Sukumar related to me, one referee for his book-length version of the study claimed that "the book set a new standard in the Cambridge series in Applied Ecology and Resource Management."[37] Sukumar's elephant work has been the basis for two books: the more scientific and scholarly *Asian Elephant* and the more popular *Elephant Days and Nights.* The foreword to this second book was written by George Schaller, proof of Sukumar's acceptance in the small circle of elite U.S. field ecologists. Sukumar's decision to publish his definitive research treatments in book form has opened him to some criticism. Some critics in India see Sukumar's focus on book-length works, which scientists usually publish only late in their careers, as an effort to avoid peer review. Although books for scholarly presses such as Cambridge are certainly reviewed by peers, Sukumar's critics imply that peer review standards for articles are more exacting. But Sukumar followed the model of scientists like Schaller or even E. O. Wilson, who have consistently preferred to present their work in a more detailed and sustained format and who have alternated between writing technical and popular works. Perhaps some of the critiques have stemmed from the real phenomenon of third-rate scholars, unable to publish in any respected journal, who nonetheless can find a book publisher willing to publish their work—perhaps even a vanity press. I suggest that Sukumar's ability to write clear and interesting prose means that his science can reach a larger audience, and he is to be commended for attempting to do so.

In any case, Sukumar has been published in journals throughout the United States and Europe.[38] On the strength of his work, he was awarded the Presidential Award of the Chicago Zoological Society in 1992 and is now the chairman of the IUCN Asian Elephant Specialist Group. He has also recently established the Asian Elephant Research and Conservation Centre (run out of his office in the CES), a public charitable trust.

Sukumar's elephant work is a good example of an ecological project that blended techniques and theories developed by Western scientists (particularly those linking behavior with evolutionary fitness, passed on to Sukumar through Gadgil) with a central research question that arose from uniquely Indian concerns. Since his project, such human–animal studies have become almost a standard at places like the Wildlife Institute of India and even some Western studies now include a section on human–animal interactions. All of Sukumar's ideas have not been received without debate, however.

Sukumar mentioned that he had submitted a paper to the U.S. journal *Conservation Biology* and was ultimately turned down. The paper suggested a pragmatic solution for crop raiding. As not all male elephants raided crops and only a few were responsible for the vast majority of crop raiding and human deaths, he suggested that those elephants be killed.

one of the referees said that I was being reactive rather than being pro-active, and that we should keep our hands off all endangered species. But the elephant is not that endangered. . . . [I]n India you still have twenty-five thousand animals. I was not arguing, Let's give the people a free license to go and take the elephants out. It really was based on the ground situation, where you find that there are one or two male elephants that are causing a lot of problems, and removing these two animals will do a lot more for the long-term conservation of the species as a whole than letting this kind of conflict continue. I really thought that this was a person sitting in an ivory tower with absolutely no idea what was going on here.

The paper was rejected by *Conservation Biology* but was eventually published in a prestigious British journal, *Biological Conservation*.[39]

Sukumar believes that *Conservation Biology* rejected his paper because of culturally different experiences with animal conservation and extinction. As he pointed out, with a far lower human density, Americans had succeeded in wiping out most large predators in the lower forty-eight states, and the reintroduction of the wolf into the nearly unpeopled landscape of Yellowstone National Park was hugely controversial. U.S. conservationists are placed in a very defensive posture when dealing with dangerous animals in proximity to people or livestock. They are used to arguing with politicians and "wise-use" groups, who argue that even the threat of wolves killing livestock is intolerable and who further argue for the right to kill wolves if found eating livestock. Conservationists in the United States thus are often put in the position of being apologists for dangerous animals that kill livestock and (very rarely) people.

In contrast, villagers in India, even those who lose crops and occasionally relatives, have coexisted in close contact with dangerous wildlife for centuries without driving those animals to extinction—before the British there is no evidence of Indian "predator elimination hunts" in the style of Texas rattlesnake drives or the turn-of-the-century wolf bounties in the United States, aimed at the eradication of every member of a given species. As Sukumar describes, animals like elephants, tigers, cobras, monkeys, and peacocks are intertwined with the religious beliefs of many Indians.[40] The danger of extinction for India's large mammals is more a result of habitat destruction, the encroachment of cultivation, and economic markets for their skins and bones (outside India) than any indigenous desire to kill them off. These different cultural contexts might lead an American scientist to fear that the boundary between killing two male elephants and killing all the elephants would be easily bridged, while an Indian ecologist like Sukumar might easily reason that if only two elephants are killed, the local populations would not take that as a license to hunt down others as well, and in fact might even feel relieved of the desire to kill other elephants in retribution.

Another Indian scientist pointed out that *Conservation Biology* did publish an article that called for culling lions. Further, many Indian scientists are used as reviewers for this journal, among others. Drawing broad conclusions about culture and science strictly on the basis of one peer review experience is dangerous—personal idiosyncrasies certainly play a role. It is still a suggestive story, but a flimsy reed, by itself, on which to build a complete argument.

Sukumar is savvy about arguments that science is a product of culture—he is related through marriage to Steve Fuller, a rhetorician of science who participated in Iowa's Project on Rhetoric of Inquiry, and is familiar with Fuller's work.[41] He accepts as obvious the contention that people choose research problems based on cultural concerns. He feels it is important that every society set its own research priorities so that they will support political or social agendas but not be compromised by them. Like Gadagkar, Sukumar believes that the techniques and methods of science are culturally neutral and can be applied to locally generated concerns and questions. He is clearly hesitant to say that science should be completely in the service of politics, with all the negative connotations that entails, but the integration of local questions with international accountability appears to be Sukumar's golden mean.

Since 1988, Sukumar has been part of an international project working to understand the dynamics of tropical forests. In 1980, Steve Hubbell, then an ecologist at the University of Iowa (since moved to Princeton and currently at the University of Georgia) began a long-term study of a fifty-hectare plot of neotropical rain forest at the Smithsonian research station on Barro Colorado Island in Panama. Hubbell and his assistants tagged and recorded every chest-high stem in that plot that was above two centimeters in diameter. The final count included 315 plant species and thousands of individual plants. The data were entered into a computer, and since then the plot has been regularly checked to record year-to-year changes. Working at an ecosystem level in the field had never been attempted before; the previous standard for large-scale forest research had been one hectare.[42]

The project attracted a great deal of global attention. On a visit to the United States, Sukumar made a side-trip to Panama to see the plot. While

there, "Steve argued convincingly why such large-scale plots were necessary in order to be able to study tropical forest dynamics effectively. I decided that we would repeat this scale of study in Mudumalai [National
Park in southern India]."[43] Hubbell was active in encouraging other scientists to set up such plots throughout the world. By 1999 there were additional fifty-hectare plots in Malaysia, Taiwan, the Philippines, Sri Lanka,
India, Thailand, Borneo, Ecuador, and Cameroon. The Smithsonian helped
to fund Hubbell's research and, through the establishment of the Centre
for Tropical Forest Research, has also helped oversee this international
network of scientists and plots. These plots can be used to generate comparative data about tropical forests throughout the planet. This project is
one of the largest ecological research projects ever undertaken, involving
vast numbers of people and funding from several different private and
governmental organizations. For Sukumar's plot in Mudumalai alone (he
chose to tag every stem over *one* cm. in diameter), 25,929 plants were
tagged and are being monitored.[44]

Sukumar has used the data from his plot not just to study forest succession but also to compare the effects of fire, human disturbance, and elephants on the forest and its species composition (the plot is large enough
so that he can see different areas being effected by different events). In this
way he has been able to use this internationally conceived project for his
own goals. He is unusual in this. From the perspective of one U.S. scientist
who has worked on the plots project, India is unique among the various
nations participating in his project because "India has this elite of scientists
who do world-class science, and they have a series of extremely highly selected graduate students who also do elite science." Sukumar has thus taken
the idea of Hubbell's plots, developed it for his own agenda, and run it
through the CES. In many of the other countries participating in the plot
project, though, "the managers and researchers have not taken a lead in ex-
ploiting their plots."[45] But the U.S. scientist was reluctant to criticize any
other research teams "because they are struggling with tremendous odds to
do this work." Usually the other nations rely upon the United States and
the Smithsonian for funding, research support, and equipment. The U.S.
scientists who lead this project recognize that not only can they not supply all these things but also that it is important to try to help build up indigenous ecologists and institutions: "We also realized that to get this done
we had to involve local scientists; that this couldn't be . . . imperialistically
driven." These U.S. scientists have resisted sending their students or U.S.
ecologists to oversee the plots. They restrict their work, for the most part,

directly to Panama and Barro Colorado (again, we see the tendency to view BCI as practically a U.S. territory—U.S. scientists never mentioned getting Panamanian scientists to take over the BCI plot). Some assistance has been given to nations setting up their plots, but the understanding is that local scientists and institutions will operate and monitor the plot over the long term. As various scientists repeated several times, though, India was completely different than any of the other cooperating nations. Sukumar and his students need no outside help. U.S. scientists coordinating the international plots see Sukumar as a colleague and have felt no need to develop Indian scientists or institutions.

Gadgil's Legacy

Thirty years ago, when Madhav and Sulochana Gadgil turned down offers to teach at Harvard, Princeton, and MIT in favor of returning to India, they could have had no idea that decision would eventually result in the formation of the CES. It was an act of supreme confidence in their ability to make a good future for themselves in India. Their confidence is at least partially attributable to the economic comfort and privilege that was also a precondition for Ali's groundbreaking work in the earlier decades of the twentieth century. It was also, however, an act of patriotism and commitment to India. The Gadgils simply did not want to abandon India and were willing to work harder, or in different ways perhaps, to help develop the scientific institutions and traditions in India that would make their desired type of research possible. They were successful.

The CES is one of the world's leading ecological institutions, their faculty conducting research that would grant them tenure in the best U.S. universities, and their students encouraged to conduct Ph.D. research equivalent to that of graduate students in the United States and Europe. There is no doubt of the close connections between the CES and the vision of modern ecology held by E. O. Wilson when he was training Gadgil in the late sixties. As such, the CES is not just Gadgil's legacy, but Wilson's as well. But Wilson could never have predicted the research questions that would engage the faculty and students of the CES or how the juxtaposition of Indian concerns and Harvard theories and methods would turn out. The CES in recent years has provided a platform for work that is at times critical of U.S. ecology. Ironically, as theoretical quantitative ecology became increasingly institutionalized at the CES, Gadgil in particular moved his research in a completely different direction. Still, the CES is also the place in India where a U.S. ecologist would be most comfortable working and conducting research.

For some critics these close connections and shared interests suggest that the CES is a product of cultural (scientific) imperialism, that a U.S.-inspired way of working and seeing the world has been grafted onto these Indian scientists. But the scientists of the CES are neither clones nor rebels—they are not merely fulfilling U.S. research agendas, nor are they bravely struggling to carve their own space out of a hegemonic U.S. scientific world. They cannot help but to be aware of U.S. theories and methods, or what passes for the hot topics of the day, but they do not let that knowledge definitively shape their own interests or techniques. Instead they are pursuing questions they find relevant and using methods that seem suited to their local needs. If those methods were developed in the United States, they do not hesitate to use them, but if American methods do not work, they modify them. If trendy American research questions are relevant to a local Indian species, they seek to answer them, but if the American questions are irrelevant, they ask their own. In so doing, and in publishing their work (whether in articles in top journals or in books from top presses), they then change the way that U.S. scientists conceive and practice ecology. The CES is far away from the publishing and funding axis of ecological power on the East Coast of the United States, but for scientists studying sociality in insects, elephant conservation, or tropical forest dynamics, to name just three areas, these Indian ecologists contribute to the dialogue that comprises the intellectual center.

CHAPTER SIX

Science to Save the Natural World

The Ecology of Conservation

> How many mountain gorillas inhabit the forested slopes of the Virunga volcanoes, along the shared borders of Zaire, Uganda, and Rwanda? How many tigers live in the Sariska reserve of northwestern India? How many individuals of the Javan rhino are protected within Ujung Kulon National Park? This many dozen gorillas, this many dozen tigers, this many dozen rhino. The numbers change marginally, year to year, situation to situation, while the recurrent questions vary also in their particulars, remaining fundamentally the same. How many left? How large is their island of habitat? Do they suffer small population risks? What are the chances that this species, this sub-species, this population will survive for another twenty years, or another hundred?
>
> —*David Quammen,* **The Song of the Dodo**

The Delhi Ridge

RUNNING THROUGH THE MIDDLE of Delhi is an elongated rise in the land people refer to as the Ridge. In several areas it is still covered with a dry acacia forest and near Delhi University a portion of it is maintained as a small park about the size of six city blocks. A wall surrounds this park with a few iron gates through which you can enter, and paved and unpaved trails run throughout it. I took several walks in the Ridge in the months I was working at the University and I usually found them relaxing. On one of these walks, I stopped to sit on a bench and watch a group of three-striped palm squirrels playing in a clearing. Palm squirrels have chipmunk-like bodies, but they add the bushy tails of a North American gray squirrel and a chirping bark that seems incongruous coming from a mammal rather than a bird. According to legend, these squirrels, found throughout India, received their stripes when they helped Lord Rama rescue Sita by building a bridge from India to Sri Lanka with their own bodies. As Lord Rama passed over them, he stroked the squirrels' backs with his hand, leaving the dark stripes that are found there to this day.

In the midst of this tableau, this physical embodiment of the Indian cultural synthesis of the natural and the sacred, a mongoose—no respecter of myth—leapt from its hiding place. With chirps of alarm, the squirrels scattered, all but one. The mongoose and the squirrel wrestled briefly, rolling in the dirt. Then the mongoose got a stranglehold on the squirrel's neck with its sharp teeth, slammed the squirrel to the ground (apparently breaking its neck), and flung the squirrel over its shoulder, all the while keeping its death-grip. The mongoose and its prey disappeared into the brush. Rose-ringed parakeets and gray and black crows were screaming in the trees above, and the noise attracted the attention of a lone rhesus macaque sitting nearby. He scrambled over, took a look, then returned to his perch in the top of a nearby dead tree as the birds calmed down. I was not nearly as nonchalant as the monkey. I could not believe that I was sitting in the middle of Delhi—the smell of rickshaw exhaust in my nose, the feeble smog-shrouded sun casting no shadows—instead of sitting in the middle of a jungle or a Discovery Channel filming site. As I watched and waited, the red-vented bulbuls returned to the ground, flitting about, and a green barbet hopped around on a low bush.

Settings such as this—little parks in the midst of human habitation—are what many Indian ecologists and environmentalists see as a model for the preservation of Indian biodiversity. Through a network of such small parks, including traditionally maintained sacred groves,[1] a wide range of India's habitat and biological diversity might be preserved. Many middle-class Delhiites are quite proud of the Ridge, and a local environmental group made a small booklet on its biodiversity: the vertebrates alone number well above a hundred species. According to these supporters of small parks, this is the solution to balancing the needs of India's biological diversity and its people. Rather than focusing on the creation of a few hundred huge national parks and removing all the people from these areas, India's environmental policy should be oriented toward thousands of smaller-scale parks, sacred groves, and patches of habitat that might have some low-level human use. Why dislocate long-established communities living within recently created national parks, they argue, when biodiversity is already being maintained in traditional arrangements?[2]

 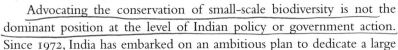Advocating the conservation of small-scale biodiversity is not the dominant position at the level of Indian policy or government action.

Since 1972, India has embarked on an ambitious plan to dedicate a large portion of the Indian subcontinent to the preservation of biodiversity. By 1975 there were five national parks and 126 sanctuaries in India. In 1997 there were 65 national parks and 425 sanctuaries, covering over 4 percent of India's landmass.[3] But these parks are almost always large—some are thousands of square kilometers—and they have been formed without reference to sacred groves or traditionally maintained refugia. As the vast majority of these parks and sanctuaries were established only within the last twenty-five years, many of their borders include preexisting human settlements—both rural villages and forest-dwelling tribal communities.[4] By law, national parks are required to be unpeopled, and though only 40 percent of the national parks currently meet this criterion, several thousand people have been relocated thus far, and thousands of others are being encouraged or forced to move.[5] The Indian government's decision since 1972 to focus on the preservation of nature in large national parks and wildlife sanctuaries rather than smaller parks and sacred groves was encouraged by Western-dominated international organizations such as the International Union for the Conservation of Nature and the World Wildlife Fund. This was a decision with a direct human impact and one justified by ecological theories of conservation policies and reserve design originating in the United States in the sixties and seventies.

Ecological Theory and Conservation

In the early seventies a growing number of ecologists began developing scientific theories and models to determine how best to save what appeared to be a rapidly degenerating natural world. They attempted both to develop and to give scientific credibility to effective global models for species and ecosystem conservation practices. This was a new focus for ecologists. Earlier scientists like Ali or Ripley had been interested in conservation. Their science, however, focused on uncovering ecological information on species or communities of species. When Ali or the editors of the *JBNHS* spoke of the need for scientific information to support their conservation initiatives, they usually were referring to counting populations—assuming that a low population would show the need for establishing a nature reserve and keeping people and their livestock out. As often as not, nature reserve policies were predicated on the knowledge derived from centuries of game management for hunting, whether in Europe, the Indian princely states, or the United States. Scientists like Ali or Ripley did not explicitly test different types of reserves.

The new conservation-oriented ecology (emerging in the late sixties) focused the efforts of the scientists directly on the efficacy of different conservation practices. It attempted to answer, for instance, whether or not the Delhi Ridge and other small parks were adequate starting points for the preservation of Indian biodiversity. This question fits into a long-standing debate among these conservation-oriented ecologists. The SLOSS debate centers around which is better—a Single Large Or Several Small reserves. To date the theories formulated by ecologists favoring a single large reserve have been dominant and have underpinned the establishment of large national parks and sanctuaries throughout the world since the early seventies—as in India.

By the late eighties, this subfield of ecology had its own name—conservation biology—and it had developed into a distinctly different branch of the science. If the CES had brought one type of Harvard ecology to India (the modern quantitative ecology of Gadgil's dissertation), conservation biology represented a second distinct trend in twentieth-century ecology, one similarly linked to Harvard and the East Coast institutions of the United States. A few scientific theories have been particularly crucial in the development of conservation biology, which has shaped the development of the Wildlife Institute of India and of conservation practices and policies of the Indian government.

Island Biogeography

Islands, with their limited size and "self-contained" ecosystems, seem to stimulate the imaginations of naturalists and scientists.[6] Naturalists have been describing and writing about the relative abundance of flora and fauna on different-sized islands for hundreds of years, as with Charles Darwin's accounts of fauna on the Galapagos Islands. In the twentieth century, ecologists attempted to move beyond these observations to a better, more theoretical, understanding of the relationship between island size, isolation, and biological diversity.[7] A pathbreaking contribution was made in 1963 when two ecologists, Robert MacArthur and E. O. Wilson, published "An Equilibrium Theory of Insular Zoogeography," which crystallized several of the ideas earlier ecologists had suggested; it also derived a mathematical equation to relate island size and species richness. Although it was certainly not produced in a vacuum, the article was original to a degree rarely seen in science—the authors cited only eleven scientists besides themselves, and of those five were referenced as the authors of statistical textbooks or plant or animal identification manuals.[8]

Four years later, MacArthur and Wilson published a book-length elaboration, *The Theory of Island Biogeography.*[9] Even today hardly a paper is published dealing with habitat fragmentation, species extinction, reserve planning, or species colonization that does not cite one of these two studies (usually the second).[10] One introductory text has even defined conservation biology as "an attempt to integrate scientific theories of island biogeography into conservation practice."[11]

In their theory MacArthur and Wilson attempted to determine how the size of islands affected the biodiversity they contained and the rate at which species became extinct, as well as how the islands' relative isolation affected their rate of recolonization from other islands or the mainland. Based on empirical studies in the Caribbean they assumed the Species-Area Curve, which states that there is a direct correlation between the size of an island and the diversity of plant and animal species it can support. They also came up with mathematical equations to determine the recolonization of islands in various degrees of isolation. Their book was riddled with mathematical equations, but the final message, as summarized in Wilson's autobiography, was relatively straightforward: "a reduction in habitat is inexorably followed by a loss of animal and plant species."[12]

The take-home message might have been simple, but there was much more in this book, and it was couched in the authority of mathematical proof.[13] Subsequent work by MacArthur and Wilson seemed to validate their hypothesis. Wilson in particular carried out a series of experiments in the Florida Keys in which he and a graduate student, Dan Simberloff, recorded the species richness of several small mangrove islands. Then they exterminated all nonfloral species by setting up a tent around each island and releasing poisonous gas—very similar to fumigating termites by tenting a house—and observed how rapidly the island repopulated with insects and other small vertebrates. Their observations coincided with their predictions, as described by the species-area curve as well as the recolonization curves.

Surprisingly, though, other attempts to test the accuracy of MacArthur and Wilson's mathematical equations were slow in coming; many ecologists were willing to accept the theory at face value. The mathematical equation at the heart of the theory, $S=CA^z$ (S is the number of species, A is the area), hinges on the existence of two "constants," C and Z, whose validity is questionable since they were not derived from theory or mathematical proofs but were devised by MacArthur and Wilson to make the equation fit their observed data from the Caribbean.[14] These shortcomings did not stop the theory from attracting fervent followers, however.

Figure 9. This herd of Nilgiri tahr walked within eight feet of me and was completely without fear. There are fewer than three thousand of these highly endangered mammals left in the world, all found in the mountains of South India (this herd is in Eravikulam National Park). Notice the patchwork of tea plantations covering the mountains in the background, outside the park boundaries.
Cassandra Kasun Lewis, May 1998

The theory of island biogeography, while pathbreaking in its own right, gained even greater importance when ecologists read *The Theory of Island Biogeography* as showing a path for nature management. That is, they made the intuitive leap to suggest that what was true of oceanic islands was also true of terrestrial nature reserves, islands of "nature" in a sea of development or agriculture. Though this theory is given credit for "first prompting biologists to consider population dynamics and extinction in relation to the size of nature reserves,"[15] Jared Diamond is generally credited with popularizing the application of island biogeography to terrestrial reserves.[16]

In a series of articles published between 1969 and 1975, Diamond both made a case for the relevance of his application of the theory and then expanded the theory's implications to include such details as shape and placement of nature reserves. The most important of his articles was a summary he wrote in 1975 that collected the "state of the art" into a single publication.[17] Diamond explicitly argued that, as island biogeography demonstrated, nature reserves needed to be as large as possible in order to preserve the greatest number of species. He added to the theory by pointing out that the shape of a reserve was also important because shape directly determined the edge-to-interior ratio and that there existed an "edge effect." The higher the ratio of edge to interior was, the greater the impact of the edge on the species composition would be. He argued that it matters, for instance, that the edge of a forest has different microclimate features than the center, with lower humidity, greater light, higher winds, and thus slightly different vegetation patterns. Edges favored certain species (plants and animals both), and thus the addition of edges would shift the species composition of the area. If a bird required a dark forest for nesting, it might need to live in an island or reserve large enough so that the edge forest did not comprise the whole island.[18]

Diamond's 1975 article was widely circulated and spawned a number of articles similar in theme. The appropriateness of moving the discussion from literal islands to metaphoric ones was seldom questioned in the U.S.[19] Although many nature reserves held more species than similarly sized oceanic islands, the assumption was that over time the terrestrial reserve would lose species until it had reached an equilibrium "equivalent to comparable oceanic islands."[20] Thus, to determine how many species a given reserve might hold over the long term (defined as one thousand years), ecologists would look for a comparable oceanic island.

Once the leap was made from oceanic to terrestrial islands, studies within island biogeography quickly devolved into the "island dilemma: how big did a reserve need to be to maintain species diversity over time?"[21] With one notable exception, ecologists did not test island biogeography itself (in the manner of Wilson and Simberloff's mangrove island experiment).[22] Instead they accepted its findings as valid and began to apply them to reserve management and planning. A text for collegiate conservation managers-in-training states that "conservation biologists have moved away from testing island biogeography theory toward two fundamental questions: How low can you go [population size] and How big is big enough?"[23] Unsurprisingly, but in accord with the finding that large islands had more species on them than small islands, many ecologists began to champion the need for large reserves to preserve maximum species diversity. Thomas Lovejoy, chief scientific advisor for WWF–USA since 1973, as well as one of the key scientists at the Smithsonian Institute and a regular expert for the U.S. Congress, writes, "Evidence from islands, as well as the theory of island biogeography, have contributed to our understanding of the problem [of size]. Since large islands hold more species than smaller ones with equivalent environments, it is reasonable to conclude that larger reserves will hold more species than smaller ones."[24] Underlying this statement is the premise that scientists can compare different-sized islands as if they were equivalent environments—and are environments ever really equivalent, throughout the world?

Minimum Viable Populations

The theory of island biogeography worked well with another rapidly developing idea used in justifying large reserves, minimum viable populations.[25] Based largely on genetics and speculation about the dire effects of inbreeding, this theory suggests that in order for a species to survive for the long term (thousands of years), there must be a minimum viable population. Several historical examples seem to support the assertion that, once they fall below a certain population threshold, many species become extinct.[26] This idea is largely untestable, however, except through computer models and best guesses, because the time scale dealt with is usually a thousand years. As a leading advocate of minimum viable population models admitted, "Intuition, common sense, and the judicious use of available data are still state of the art." Thus, "minimum viable populations on the order of a few hundred to several thousand genetically effective individuals are within the range that satisfy those scientists who have attempted to deal with real management situations."[27] This was apparently said with no irony, but there is a tremendous practical difference between having to save seven hundred or seven thousand tigers in order to preserve the species.

This theory fit in with island biogeography in its emphasis on the need for nature reserves that are "as large as possible, and there should be many of them."[28] According to minimum population considerations, conservation managers should multiply the minimum viable population figure by the range requirements of a single animal to obtain what is called the minimum size of a viable reserve. (Many radio telemetry studies of free-ranging large mammals, such as Williams's study of elephants, are implicitly oriented toward collecting data to be used for this type of calculation.) For example, this theory would suggest that in order to preserve the North American mountain lion with a population of five hundred animals (commonly—but somewhat arbitrarily—used as a benchmark figure),[29] a reserve of at least thirteen thousand square kilometers would be required (about the size of Connecticut). This calculation is highly arbitrary, though, for both numbers are easily called into question. The scientists working on minimum viable populations admit that their best numbers are educated guesses. Similarly, how reliable are home range estimates? If the home ranges of carnivores are dependent on the density of prey species, for instance, how can that yield a reliable range size for areas with different numbers of prey species? Nonetheless, calculations of minimum viable populations, and the corresponding minimum viable reserve size, became widely accepted as scientifically valid.

Michael Soule and Daniel Simberloff were aware of the problems with minimum viable population figure, but, they concluded, "there are good reasons for suggesting scientifically based guidelines and standards for the practice of conservation. Without them, pro-conservation individuals and groups, in and out of governments, hardly have a leg to stand on when competing for land and resources with powerful elements arguing for appealing, short-term or ill-conceived development activities."[30] Both scientists knew they were supporting environmental advocacy with misleading scientific language and theory to strengthen their position. The crisis of environmental destruction seemed to warrant this, however. The alternative seemed to be to give up the fight to those who would develop every natural area, clear-cut every forest—and is it not better to err on the side of caution? In the mid-seventies, Simberloff was driving to the Keys to survey some field sites. On his way he passed some of the original islands he had studied as a graduate student with Wilson. One had been cleared of trees and converted to a trailer park.[31] In the same way that some ecologists used minimum viable populations to give credibility to their conservation agenda, island biogeography had made "common sense" about the need for large reserves into mathematically supported ecology. It seems obvious that larger islands or reserves hold more species than small ones, but obviousness alone makes for poor scientific proof in a world in which it is obvious that the sun circles the earth every day.

The Debate over Reserve Size: SLOSS

There were ecologists who challenged the assertion that large reserves would necessarily contain a greater number of species. Although most ecologists supported the idea that nature reserves should be as large as possible, there were still several who believed that "many theoretical underpinnings in island biogeographic theory have been inadequately tested."[32] Ironically, the first significant paper to publicly attack the dominant single-large-reserve model was coauthored by Daniel Simberloff, the graduate student of Wilson who had later worked with Soule on minimum viable populations. Although Simberloff supported island biogeography, he objected to its uncritical acceptance and application in situations for which it was not designed or tested.[33] He also specifically refuted the general statement that "island biogeography dictates that single large reserves are generally preferable to groups of small ones."[34] Simberloff was worried that governments who read the new ecological arguments for large reserves would give up on the preservation of small reserves. David Quammen quoted Simberloff: "[Israelis] read, both the military and agricultural interests, that small reserves are not able to preserve viable populations. And almost all the reserves in Israel are very small. So at a cabinet-level meeting they said that the weight of scientific evidence shows that these are worthless anyway, and what's the good of them?"[35] The publication of his paper in 1976 marked the official beginning of the SLOSS debates, which continue to rage through ecological journals.

Proponents of the several-small-reserves position argued that, if you had a hundred square kilometers to devote to a national park, you might be better off with ten parks of ten square kilometers each than one big park. By having several small parks spread across a larger total area, rather than a single large one, you gain several advantages: you can probably cover more microhabitats (meaning a greater diversity of land types and ecosystems), you are not putting all your biological eggs in one basket (in the event of disease or fire), and if the parks are sited somewhat close together or have linking corridors, the animals can still move around and recolonize other small parks that might have suffered local extinctions (as per the theory of island biogeography). Although small parks might not accommodate large mammals, several small parks would almost certainly guarantee that a greater number of species of insects and plants would be preserved.[36] Given that even supporters of the single-large position agreed that "species extinction of the large vertebrates seems inevitable,"[37] why should the most effort be put into preserving those species through large reserves?

Although the debate has been heated, actual comparative studies have been slow in yielding results. In 1979, Thomas Lovejoy persuaded the government of Brazil to allow him and a group of colleagues to preserve fragments of various sizes—from half a hectare to several square kilometers—out of the Amazonian rain forest as loggers clear-cut around them. (Brazilian law required 50 percent of any land scheduled for logging be preserved, and Lovejoy persuaded Brazilian loggers and officials to preserve land in a series of experimental plots of increasing size.) His team has been watching those forest fragments since 1980, checking to see how rapidly species diversity falls, if at all.[38] More locally, at the University of Iowa, Steve Hendrix has been comparing large and small prairie fragments for over a decade, monitoring plant and insect biodiversity. By 1986, "virtually all conservation biologists agree that several small reserves can contain as many species as a single large one at the time of demarcation from their natural surroundings."[39] There is no agreement, though, on what happens in the years following habitat fragmentation, whether or not small reserves can sustain their diversity, and the value of the different sorts of species saved—is it more important to save many species or one very important species? This last question, of course, is a question of values, not of science. Scientists who study large mammals, or conservation organizations oriented toward charismatic species such as elephants, pandas, or tigers, usually support the single-large model and argue that if you save the big species, the smaller ones will be protected as well (the umbrella species model). Scientists who study plants, amphibians, or insects often support the several-small model, unless they are interested in large-scale nature conservation for its own sake, in which case they might support the single-large position.

After over twenty years of argument, many scientists admit that SLOSS is an unanswerable debate. Both single large and several small are appropriate in different contexts, and for different goals. Any ecologist would admit that a huge park does not have any inherent ecological *disadvantages,* but such a park is often not politically or socially preferable, let alone feasible. And in contrast, even supporters of large parks admit that more total species are saved by carefully placed small parks in a variety of habitats but that such parks cannot sustain larger animals, as well as many of the nongeneralist smaller species. And *nobody* would want Israel to give up on its small reserves or America to give up its large ones.

People in Nature

What is often left unsaid is that the SLOSS debate masks a second debate over a central assumption of conservation-oriented ecology—is it sufficient to simply preserve biodiversity, or must pristine, unpeopled, naturally functioning ecosystems also be the goal? Although small reserves often maintain a high biodiversity, they sometimes have to be managed in an unnatural fashion (for instance, culling certain animals when the population becomes too high, as with elk in Yellowstone National Park, or pulling up plants that are growing too quickly in a wetland). Proponents of large reserves have seized on this aggressive management as a defect. Many ecologists are strongly committed to the idea that naturally functioning ecosystems are worth saving in and of themselves and that any human involvement in any ecosystem is, by definition, disturbance from the normal or "natural" workings of that system. Among other problems, a human-managed ecosystem will no longer allow the invisible hand (my phrase, not theirs) of evolution to work freely in shaping potential new species, these ecologists argue. Arguing for large reserves is arguing for ecosystems of a sufficient size so that human management is not needed.

This position is closely linked to the idea that *nature* means the absence of people. Just as ecologists searched for unpeopled wilderness in which to practice their science, ecologists advocated the preservation of nature in unpeopled reserves. This idea, that any human involvement in any ecosystem is, by definition, disturbance from the normal or "natural" workings of that system, has a long cultural history in the United States.[40] Implicit in the several-small position is the idea of small nature parks scattered in a matrix of human use. This is anathema to the scientist who sees any human contact with the natural world (short of science, we might assume) as harmful.

In making the case for a large nature reserve that does not need human micromanagement and does not include people in its midst, ecologists are acting on a vision of nature that is culturally specific. It also sets a standard for autonomous nature reserves that may be impossibly high for nations that do not share the wealth and low population density of the United States. Embedded in the midst of their argument for minimum viable populations, Soule and Simberloff write, "the Everglades in southern Florida are threatened by diversion of water that is occurring outside the park. Wherever possible, reserves should contain complete watersheds."[41] What they are obliquely suggesting is a park that extends from directly south of Orlando all the way to the tip of Florida, including the Kissimmee River, Lake Okeechobee, and the wetlands south of the lake. Although this sounds radical, the Army Corps of Engineers is currently in the process of purchasing the land surrounding the Kissimmee so that it can once more flood and ebb as it once did prior to channelization, and the state of Florida is buying as much of the land between Lake Okeechobee and the Everglades National Park as is possible. It is clearly the goal of government policy to bring about exactly what Soule and Simberloff suggested, the protection of an entire watershed. On October 19, 2000, Congress agreed to appropriate 7.8 billion dollars to the restoration of the Everglades water supply.[42]

The other solution, of course, is for ecologists to actively manage the flow of water into Everglades National Park to approximate traditional water levels. This solution, though, is thought to be a poor second choice. Is it not hubris, after all, to think that people can know how to mimic nature? As attractive as the Everglades project appears, however, it may be an unreasonable model to push on other parts of the world. The eight-billion-dollar price tag does not even include the earlier work approved to reconstruct the Kissimmee River or the purchase of agricultural land by the state.

In any case, single-large supporters have carried the day at the level of advocacy and policy. Conservation biology was defined in a 1992 text as "a friendly, mission-oriented science that justifies the necessity for large areas of inter-connected wilderness." Among environmental NGOs the calls for large reserves ranged from the mild ("from both Island Biogeographic theory and the theory of Minimum Critical Size it is evident that very large areas must be protected") to the utopian (as when two ecologists argued that in planning a successful reserve one must take into account the largest historical disturbance to that ecosystem—a fire, for instance—and then make the reserve "fifty to a hundred times larger"). A writer for the WWF, in a book on conflict resolution in the implementation of protection measures, declared that "it is generally better to address larger areas than smaller ones." The World Conservation Strategy, published in 1980 by the IUCN, provided its one line answer to the SLOSS debate: "single large refuges are generally preferable." Ecologists arguing for several small reserves have been marginalized within ecological journals, and completely ignored at the level of public policy recommendations, left to fume, as two wrote, about the "uncritical acceptance and application of an insufficiently validated hypothesis."[43]

The marginalization of an otherwise scientifically sound position (several-small) resulted from the monopoly of advocacy and advisory positions exercised by the supporters of the single-large position. At the level of policy, the SLOSS debate was not decided on the basis of competing rhetoric—the marshaling of evidence to construct an argument persuasive to the other side or that invalidates the contrary position—but on the basis of one side corralling the institutional power to define terms and make policy recommendations. The last thirty years, besides seeing the SLOSS debate, also witnessed the deliberate movement of a group of ecologists into environmental policy advocacy. When these ecologists assumed positions of leadership within the WWF, IUCN, and the Smithsonian Institution they supported the single-large model, and they pushed to see that this ecological theory was accepted as orthodoxy.

The Rain Forest Mafia

Many ecologists from both sides of the world claim that there is a separation between science and advocacy and, correspondingly, between ecology and environmentalism.[44] This has traditionally been the ecology party line: ecologists do neutral science, and environmentalists argue for a position. Many ecologists told me that they bend over backward to avoid making statements that could be construed as environmentalist because they do not want to impinge on the validity attributed to their science. Wilson writes in his autobiography that "as the 1970s passed I wondered, at what point should scientists become activists?" Wilson states, both in his autobiography and in person, that when he and MacArthur created the theory of island biogeography, they did not consider its conservation applications. Supporting this is the fact that scientists such as Jared Diamond and Dan Janzen, not Wilson or MacArthur, expounded the conservation implications of island biogeography in print. In fact, in the years following the publication of *The Theory of Island Biogeography,* Wilson turned his attention to his work on social insects, culminating in *The Insect Societies,* a text still viewed as an authority on the subject (regarded as the bible in certain entomology labs), and then in the mid-seventies, to his most (in)famous book, *Sociobiology.* He claims that "the years writing the two syntheses were among the happiest of my life."[45] They were not years of environmental advocacy.

Other biologists *had* begun to actively work in the public sphere, including Rachel Carson, with *Silent Spring* (1962), and bird ecologist Paul Ehrlich, with *The Population Bomb* (1968). Wildlife biologists like George Schaller, Sálim Ali, and Dillon Ripley actively argued for conservation throughout the sixties. Beyond ecology, the sixties saw a phenomenal growth in the U.S. environmental movement, from the passage of the Wilderness Act in 1964 to the growing concern over environmental degradation and endangered species that culminated in the flurry of environmental legislation in the early seventies.[46] In 1964 the International Council for Scientific Union[47] started a ten-year project, the International Biological Program (IBP), oriented around environmental sciences. This project provided ecologists with ample funding for large-scale ecological studies and validated ecological studies oriented toward environmental problems.[48]

This project was one of many attempts after World War II to foster an international community of scientists but was the first to involve biologists. It had been difficult to get biologists to agree on what the IBP might be centered around—invariably, scientists had suggested their own areas of expertise as the focal field. Thus, one scientist suggested a ten-year international emphasis on nucleic acids, another on photosynthesis. As might be expected, global enthusiasm for these proposals was underwhelming.[49]

Finally, the decision was made to focus on human-environment interactions, a more topical and less specific theme. The IBP was then subdivided into seven sections, and one was called terrestrial conservation.

This terrestrial conservation subgroup was dominated by scientists already linked to each other through the IUCN, many of them wildlife biologists, and they quickly overshadowed the other subgroups with their organization and commitment to a common agenda—turning ecology to the service of nature conservation. When funding resources were allocated, this group received the bulk of the money, especially within the United States, which "was by far the world's largest [participant]."[50] The terrestrial conservation group published a series of working papers and worked with other international organizations, such as the WWF and UNESCO, to raise the profile of environmental concerns. On the basis of the early work of the conservation subgroup, UNESCO convened a conference of scientists in 1968 whose primary proposal was that "a plan for an international and inter-disciplinary program on the rational utilization and conservation of the resources of the biosphere be prepared for the good of mankind."[51]

The Theory of Island Biogeography was published in this context of rapid growth both in environmentalism and in ecologists interested in using their science to pursue environmental goals. The first ecologists who supported the application of island biogeography to conservation were already advocates of large reserves, before there even was a SLOSS debate (which began only in the mid-seventies with the advocacy of some ecologists for the several-small position). Before 1975 island biogeography was understood to advocate large reserves, and this use of the theory both justified and solidified previously existing practice.

In 1959, UNESCO had asked the IUCN for a list of the world's national parks and equivalent areas and the criteria for them (the chair of the IUCN's International Commission on National Parks was Harold Coolidge of Harvard University) and when the resulting list was published in 1967, "an area which is too small [was] not included."[52] Specific guidelines were listed, based on national population densities. For nations with a population above fifty people per square kilometer, a national park or protected area was required to be above five hundred hectares. For a nation with a lower density, the reserve size should be above two thousand hectares (26). This report was published in the same year as *The Theory of Island Biogeography*. When the IUCN report was updated in 1972, now with the benefit of this theory, the IUCN reported:

In general the term national park should not be applied to protected areas of small size. . . . [I]t is recommended that parks be of sufficient extent to protect not only natural vegetational diversity but also the more mobile animals [this is the footprint of arguments based on the range size of large animals] that form part of a given biotic community. In the minimum limits set by IUCN's International Commission on National Parks for national parks to be included in the UN list, a differential was previously allowed on the basis of the population density of a country. In 1972, however, the ICNP established a minimum size of 1,000 hectares to apply to all countries (with the possible exception of small islands). This means that no general exemption, covering the automatic acceptability of smaller areas, is made for those countries that have failed as yet to find a solution for their human population problems. (36)

The ecologists of the IUCN, charged with the mission of defining acceptable nature conservation policies, were clearly influenced by island biogeography (the reserve must be large) as well as by their colleague Ehrlich's 1968 critique of overpopulation (there are too many people!).

Although some ecologists had begun to work on conservation-oriented ecology, many still feared that "in so doing, biologists jeopardize the societal trust that allows them to speak for nature in the first place." As the SLOSS wars raged around him in the latter half of the seventies, Wilson looked down at his ants and took "some relief from the knowledge that non-academic organizations were already active in the conservation of biological diversity." Wilson's nonchalance changed (as did that of other similarly ambivalent ecologists) with the 1979 publication by British ecologist Norman Myers of *The Sinking Ark: A New Look at the Problem of Disappearing Species.* Myers took a familiar theme for ecologists—extinction—and made it into a crisis. Wilson's theory of island biogeography was based on extinction rates, but it was a theory. Myers "published the first estimates of the rate of destruction of tropical rain forests," as it was occurring in the present. Globally, 1 percent of the rain forest was disappearing every year, Myers claimed, and one million species would be lost between 1975 and 2000.[53]

As Wilson remembers, "this piece of bad news immediately caught the attention of conservationists around the world." In 1980 environmentalists successfully lobbied the Congress to amend the Foreign Assistance Act, requiring that any program funded by the Agency for International Development conduct an environmental impact assessment. As this amendment was explained in an AID memorandum, "the destruction of humid tropical forests is one of the most important environmental issues for the remainder of this century . . . [in part because] they are essential to the survival of vast numbers of species of plants and animals." U.S. ecologists had a long-standing appreciation for tropical forests and work on tropical ecology—their remote wilderness frontiers and the cradles of evolution. When Myers wrote of the loss of the world's rain forests, he was writing of the loss of Wilson's, and many other ecologists, favorite biota. Wilson became an environmentalist and he joined what he "jokingly called the 'rain forest mafia,'" including himself, Jared Diamond, Paul Ehrlich, Norman Myers (the only non-American), Thomas Lovejoy, Daniel Janzen, Thomas Eisner, and Peter Raven—single-large supporters all.[54]

The Hazy Art of Predicting Extinction Rates

After 1979 efforts to predict global extinction rates became almost a cottage industry. Thomas Lovejoy made his intervention (1980), arguing that in the next twenty years 15 to 20 percent of the earth's species would be lost. Paul and Ann Ehrlich published *Extinction: The Causes and Consequences of the Disappearance of Species* (1981). Peter Raven estimated that two thousand plant species per year were being lost in the tropics and subtropics (1987), and Wilson estimated that there would be a 2 to 3 percent global loss of species per year.[55] These are just the efforts by members of the rain forest mafia.

Calculating extinction rates, like calculating minimum viable populations, depends on a series of assumptions. The most basic, which is clearly true, is that there are species of flora and fauna that have not yet been recorded. The second key assumption, which is probably true, is that most undescribed species are in the tropical and semitropical zone (rain forests, especially, but also coral reefs). From this somewhat stable foundation, edifices of sometimes questionable logic are built in the attempt to know how many unknown species there are in the world and how fast these unknown species are going extinct. In a 1995 summary of extinction estimates (presented without irony) in the U.S. flagship general-science journal, *Science,* the following explanations are given for how ecologists determine estimated species numbers for insects:

> A large sample of canopy-dwelling beetles from one species of tropical trees had 163 species specific to it. There are 5×10^4 tree species, and so $163 \times 5 \times 10^4 = 8 \times 10^6$ species of canopy beetles. Because 40% of described insects are

beetles, the total number of canopy insects is 2×10^7. Adding half that number for arthropod species on the ground gives a grand total of 3×10^7. . . . If only 20% of canopy insects are beetles, but there are at least as many ground as canopy species, then the grand total is 8×10^7.[56]

With their predictions written in scientific notation, their two estimates seem closer at first—a difference of only 3×10^7 versus 8×10^7. But of course those numbers are thirty million species versus eighty million. There is a bit of wiggle room in the estimates, especially considering that, as of 1985, only 1.7 million species had been identified, including roughly three-quarters of a million insects.

The whole exercise of estimating the number of insect species, though gallantly attempted, is based on too many unknowns. In this case, the researchers started with one known fact, derived from Terry Erwin's research fumigating tropical trees to count the canopy insects. He found that there were 163 species of beetles specific to one tropical tree species. From this observation, the following assumptions are made: (1) every tree species has the same number of specific beetle species not found on other trees, (2) the number of tree species is known, (3) there is no variation between the number of insect species on tropical or temperate trees, and (4) beetles truly comprise 40 percent of insect species, and this current ratio is not an artifact of what has been described (i.e., perhaps beetles have been more studied than other insects). Further, the relationship between the number of insect species in tree canopies and the number of species on the forest floor is unknown, and guesswork was used to move from total canopy species to total insect species, floor included (thus the two dramatically different estimates for total insect species). Once an ecologist arrived at a working estimate of the total number of species in the world, she could then combine that with estimates of habitat lost and island biogeography considerations concerning size to arrive at an extinction rate.

Obviously, these extinction estimates are highly unreliable, but surprisingly they were being made, and widely, by highly respected ecologists after 1979. The estimates were designed to shock and to create a climate for conservation. Why guess thirty million insect species when eighty million seems equally likely? It is undeniable that extinctions are occurring in the world and that the vast majority are human induced. Ecologists felt this loss strongly, and many, like Wilson, were pushed to enter the sphere of environmental advocacy. As David Takacs writes in his history of the move from ecology to environmentalism by U.S. ecologists, "An overwhelming sense of love [for the tropics and its diversity] and a foreboding sense of crisis lead them to redefine what it means to be a scientist; they do so to save the source of their work, the fount of their professional, emotional, and perhaps, genetic [if they agree with Wilson's biophilia hypothesis], sustenance."[57]

Biodiversity and Conservation Biology

Against this background of death and destruction, two important events occurred in 1986, two political acts that represented the formal maturity of conservation-oriented ecology. While planning a national conference sponsored by the Smithsonian Institution and the National Academy of Sciences to be held in that year, Walter Rosen (of the National Academy) suggested to E. O. Wilson that they coin a new word to use at the conference, *biodiversity*. Wilson resisted at first, thinking the word was too flashy, but he relented. The National Forum on BioDiversity, and the resulting proceedings edited by Wilson, made *biodiversity* a household word.[58] Janzen (who was invited to speak at the conference by Wilson) remembers, "That was an explicit political event, explicitly designed to make Congress aware of this complexity of species that we're losing."[59] There is no question that the conference worked. Around fourteen thousand people attended the forum, which had sixty presenters. The last day of the conference was beamed by satellite across the world and watched by thousands of scientists and students. The book, *BioDiversity,* was a best seller and the new word immediately became a widely recognized code for conservation issues.[60]

Nineteen eighty-six also marked the founding of the Society for Conservation Biology (SCB), formally recognizing conservation-oriented ecology. One of the founders of the SCB described conservation biology as "the application of science to conservation problems, [addressing] the biology of species, communities, and ecosystems that are perturbed, either directly or indirectly, by human activities or other agents. Its goal is to provide principles and tools for preserving biological diversity."[61] Although the subdiscipline was formalized in 1986, people had practiced conservation biology much earlier; a key theme of this book is the development of exactly that tradition in the United States and India.[62] As ecology emerged from taxonomical zoology and botany, the world's first national park was being established in Yellowstone (1872). Ecology and environmentalism (specifically the use of nature reserves to save species and ecosystems) developed side by side through the twentieth century, and there was considerable interplay between the two, particularly in the United States (where the SCB is based, though its membership is international). Although conservation biology gave ecologists a new platform from which to claim that "[b]iologists can help increase the efficacy of wildland management; biologists can improve the survival odds of species in jeopardy; biologists can help mitigate technological impacts,"[63] the underlying goals of species and ecosystem conservation were scarcely different than those being advocated by Sálim Ali or George Schaller (to name just two scientists) decades earlier. The naming of conservation biology, and the association created in its name, was the final political step in moving ecological science into the service of conservation goals.

The entry of ecologists, of conservation biologists, into public discussions of environmental policy has provided a scientific basis for, and sometimes determined, how international conservation organizations made funding decisions and policy recommendations in the developing world. There is little question that they have been successful. As Takacs writes, they "speak for it [biodiversity] in Congress and on the *Tonight Show.* They whisper it in the ears of foreign leaders. . . . They transport ten percent of the U.S. Senate to spend nights in the heart of the Amazon . . . [and] they weave sensuous word tapestries in books meant to seduce readers to love biodiversity."[64] These scientists, the rain forest mafia and their followers, are determined to remake the world in a more biodiversity-friendly image. And when I say "remake the world," I am not speaking metaphorically. The rain forest mafia and their supporters expressly rejected the assumption that, as one Australian ecologist wrote, "debates about reserve design and SLOSS are of limited relevance because, with very few exceptions. . . . Conservation managers . . . never have the opportunity to design a reserve network before an area is fragmented."[65] In opposition to this practical but limiting argument, the members of the rain forest mafia and their many supporters suggest that just because there are people or developments on land slated for parks does not mean they have to stay there. People can be moved; land can be rehabilitated. Nature can be restored.

Guanacaste National Park and Biosphere Reserve, in Costa Rica, was at one time a mixed farming and forest area. Daniel Janzen, professor at the University of Pennsylvania, with the help of U.S. funds and the blessing of the Costa Rican government, has converted that settled land into a national park of twelve hundred square kilometers. Janzen now lives there for much of the year, monitoring its rehabilitation.[66] If it can be done there, it can be done elsewhere; it *must* be done elsewhere—this is the political import of the ecological theories and environmental advocacy of conservation biology as it emerged in the eighties. Some people can read this last sentence and feel a profound sense of relief that the world is not irredeemable. Although I also worry about extinction and habitat loss, I fear that conservation in much of the developing world is not so simple as some displaced peasants, some money, a scientist, and a willing government.

CHAPTER SEVEN

Indian Science for Indian Conservation

Nationalism and Wildlife Biology

> [I]t is nice to have Schaller coming and doing some work, . . . but it doesn't really strike across to the rest of the Indians as, That is something doable by us. . . . Johnsingh's work . . . made things more real. . . . It was like, say, an Englishman taking a walk in the middle of the day—that is a very odd thing to do for most Indians. That is the way it seemed wildlife research was, because it was all white-skinned, pale people coming and doing it. But the time had come for an Indian to do wildlife biology; I think it was the whole generation, Johnsingh to begin with, then Sukumar, and then the real explosion of people.
>
> —*Ravi Chellam*

WHEN NIGHT FALLS AT THE WILDLIFE INSTITUTE of India outside Dehradun and the faculty have gone to their homes and offices have closed, the old hostel mess (for graduate students) is the place to be. After a few weeks the novelty wears off the food—invariably rice, chapatis (like tortillas), dal (lentils), and subji (various vegetables boiled until mushy). The conversation, though, is unsurpassed.

The graduate students of the WII are an extraordinary collection of bright, motivated, nature-loving people. They come from every region of India, yet they are more alike than different. They are supremely fluent in English, having been educated in that language. All of them, men and women, wear blue jeans, and the social room fills up every Wednesday night for *The X-Files* or the Discovery Channel after supper. Nearly all the students come from comfortable economic backgrounds, and all have excelled as undergraduates (or they would never have made it through a rigorous national screening process, including multiple examinations and interviews, to come to the WII). It seems likely that most could have succeeded in other more lucrative fields (most directly for these biology students, in medicine), and many have families that do not quite understand why sons or daughters have turned down medical school to live in relative poverty in a jungle. These students are the next generation of India's conservation biologists and policy planners, the avant-garde of India's conservation elite.

There are more reasons to come to the WII than the dinnertime conversation. The institute has been the Indian government's chosen wildlife management training and research organization since 1982. In support of the WII's research aim, the faculty and graduate students work all over India, from the coral reefs of the Andaman Islands to the highest peaks of the Himalayas, and they have unparalleled access to India's protected areas and endangered species. To fulfill its training goals, the WII runs in-service wildlife management courses for the Indian Forest Service (there is a nine-month and a three-month course every year) and is responsible for training foresters in the latest theories and practices of protected area management. The WII has been steadily supported by grants from the FAO and UNDP of the United Nations, by the USFWS, USFS, and Ford Foundation, and by a continuous stream of smaller grants from the United States. The eighty-hectare main campus is stunning, set in the countryside a few kilometers outside the town of Dehradun and facing the front range of the Himalayas that rise above the sal trees. The pyramidal rooflines of the buildings emulate the mountains in whose shadows they sit. Thanks to its international and governmental funding the campus is well equipped with networked computers, a Global Information Systems lab with mapping equipment, an impressive library with a computerized indexing system, audiovisual equipment, and many labs and seminar rooms.

The WII is where the science of conservation biology and the Indian Forest Service (IFS) intersect. It merges the tradition of conservation-oriented ecology that runs through Sálim Ali and Madhav Gadgil with the traditions of nationalism, park management, and conservation that had developed within the IFS and the governmental bureaucracy. The need for such an institute was spurred by the freeze on the funding of U.S. ecologists in the early seventies, in the context of strained Indo-U.S. relations and the fear of inappropriate U.S. motivations for ecological studies (as with MAPS and the U.S. Army). As with the BNHS and the CES, the story of the WII is best told in relation to a few pivotal figures, but unlike them (both largely the product of individual effort), the WII combines differing traditions and goals and requires a larger cast of characters. And just as Sálim Ali introduced the ecology tradition to India and Madhav

Gadgil brought the quantitative evolutionary-ecology synthesis from Harvard, so the WII has become the primary center for Indian conservation biology, the applied subfield of ecology oriented toward the conservation of species and natural ecosystems that emerged in the wake of the environmental movement of the sixties and seventies.[1]

Nationalism and Indian Ecology

Following the India–Pakistan War of 1971 and the government of India's decision to limit the access of U.S. ecologists and funds to India, some Indian foresters, bureaucrats, and environmentalists saw an opportunity to assert greater Indian control over the booming conservation movement in India and to insure that Indians developed expertise in this area and were not always dependent on U.S. scientists and "experts." The sixties had been a period of greatly enhanced concern for conservation in India, exemplified by the writings of the BNHS (funded by, and collaborating with, U.S. experts), the formation of the Indian chapter of the World Wildlife Fund (supported by the international WWF), the 1969 IUCN conference (again, heavily supported by outside experts), and the first inklings of what would become Project Tiger. The early seventies would prove to be a watershed moment in the history of Indian conservation, culminating in the passage of the 1972 Wildlife (Protection) Act (establishing a legal basis for national parks and sanctuaries) and the establishment of Project Tiger in 1973. If the sixties had represented the ascendancy of scientists in Indian conservation, the seventies was largely the decade of the government officials and the IFS—a transition made possible in large part by the exclusion of the U.S. collaborators and funds that had given nongovernmental Indian scientists credibility, access to resources and equipment, and training. The simultaneous shut-down of the flagship BNHS project (the bird-banding project) due to the governmental investigation into Ali's ties to the U.S. military helped to weaken the credibility and energy of the BNHS in the early seventies.

    M. K. Ranjitsinh, now with WWF-India, was at the center of these changes. He was director of wildlife in the Ministry of Agriculture from 1971 to 1976 and one of the most important administrators in the Indian government in the early seventies (this was before a Ministry of Environment and Forests had been established, and both the IFS and all wildlife concerns were included in the Ministry of Agriculture, which was staffed by both foresters and nonforesters like Ranjitsinh).[2] A retired administrator who had participated in this process spoke directly about why he and his colleagues had tightened research access to India in the early seventies:

> We didn't have our own boys doing research. We didn't have the Wildlife Institute. . . . People would come [from outside India], research institutes, individuals, and say, "We would like to help you." Fine. "These are our chaps, they have this degree, and they would like to do research here." Still fine. Then they would come to the basic part: "These are the subject areas which our boys or girls would like to research. . . . Why don't you take them on and why don't you see if you can help them." And I would say, "If you want to help, then ask us what we need as far as the research is concerned. We do know something of what we need." As I said earlier, you do what you want to, the breeding biology of the dung beetle, or anything you want to do, but don't say that we need it. . . . Because we know what we need. . . . If you are prepared to take on our projects, or variations of these, that would be helpful. Then you can tell the world that we need it. Otherwise don't tell the world. . . . Ultimately people did come to terms, but I remember a lot of people didn't like it . . . because they thought they could come and do research and tell us, "You poor blokes, you don't know what you're all about," and we are telling you, "This is what you should be doing."[3]

There is no doubt that some U.S. scientists took a "BCI mentality" to India.[4] They treated India as a field site, tried to ignore its people, and viewed the Indian government as a superfluous bureaucracy.[5] One U.S. ecologist described his work in India in the eighties, collaborating with the BNHS, "They were the nominal Indian collaborators. It was relatively nominal in that it turned out to be very difficult for us to integrate ourselves into their way of doing things, and vice versa. . . . Basically the whole way of doing things, the whole expectation of how to behave, and how to do work, was completely different between the two groups. . . . We worked on our stuff, and they worked on their stuff, and we talked with each other, but that was about it."[6] He and his colleagues had even refused to use the BNHS field station, renting their own house in town. He has no plans to return to India. I do not have similar evidence of poor relationships between U.S. and Indian scientists in the sixties or seventies, but that of course does not mean that they did not occur.

U.S. ecologists who were working well with their Indian colleagues would still make grandiose policy pronouncements to the Indian media, or in meetings with high-ranking government officials. U.S. ecologists like Ripley, or international conservation organizations like the WWF, would call for more research on the rhino, or tiger, or Siberian crane, or they might call for the removal of people and cattle from national parks. Often these scientists were working in conjunction with the BNHS, or in the cases of Spillett, Ripley, and Schaller, at the request of the BNHS. The BNHS would use the U.S. scientists to gain credibility and support for its projects within India. This "cult of the international expert" backfired on the BNHS and U.S. ecologists when the international experts came to look not like benevolent founts of wisdom but instead like the point in the U.S. imperialist wedge.

Hopefully you are persuaded that Indian ecologists like Ali and the BNHS (or Gadgil and the CES, though that institute was not yet formed in the early seventies) did not experience U.S. ecologists as imperialistic, especially when compared with the British, who had so recently left India. Ali knew that when Ripley made statements about India's wildlife, he often was acting at the request of the BNHS, to further their agenda. And on subjects that extended beyond their personal expertise, many Indian ecologists trusted other ecologists, regardless of where they were from, more than they trusted Indian nonecologists. Other Indians, including Ranjitsinh and his colleagues in the wildlife branch of the Ministry of Agriculture, did not see U.S. ecologists, or Ripley, in that light. How were they to know that Ripley was simply saying what Ali told him to say? They saw Americans who were arrogantly telling the Government of India what to do—Indian ecologists who agreed with the U.S. scientists were already corrupted by the siren call of U.S. cultural imperialism, the government officials might claim.[7] This resentment of foreign intervention was especially strong when the BNHS and international conservation organizations turned their attention to India's reserved forests and protected areas (the precursors to India's national parks) and suggested changes in management. This was a direct infringement upon what the IFS (who managed these lands) saw as their exclusive domain. It is likely that the IFS would have been equally vehement in opposing Indian ecologists in this arena, even without their international connections.

This change in attitude toward international experts had begun before the 1972 freeze. In 1970, Zafar Futehally (a relative of Ali's and an official in the WWF-India) was appointed to India's Expert Committee on National Parks and Sanctuaries, which was to submit a report to the Indian Board of Wildlife. Futehally promptly asked Lee Talbot of the Smithsonian for his assistance in organizing and compiling international advice from the IUCN, Smithsonian, and WWF-International. Futehally added a cautious note: "You will realize the importance of handling this affair rather carefully, for if it becomes known that foreigners are trying to tell us what to do and what not to do in our Sanctuaries, it might be resented."[8] The Indo-U.S. politics of 1971 during the India-Pakistan War did not create the dislike for foreign ecologists, it just made it easier for this perspective to move into ascendancy.

The Ministry of Agriculture officials and the IFS understood the need for scientific information, but they wanted to develop local expertise. Local expertise, for the IFS, usually meant internal agency expertise. Ranjitsinh commented, "We didn't have people [to do ecological work] in this country, and I wanted to encourage people of Indian origin to move up and take on this work." The wildlife administrators in the government of India hoped that cutting back on the number of U.S. ecologists in India would help develop Indian expertise in Indian ecology by (literally) clearing the field of competition. When Schaller tried to come back and do work in the Indian Himalayas in the seventies, he was denied permission and instead went to Pakistan, China, and Nepal. When (by then Dr.) Juan Spillett put together a research team, and obtained the funding for the project from National Geographic and the Smithsonian to do work in India, he was denied permission. The Smithsonian tried their best on his behalf, but after two years of letters and meetings in India, he finally gave up—his research couldn't be done in any other part of South Asia, so it was never done.[9] When Thomas Foose from the University of Chicago tried to do work on the rhinos of Kazaringa, he was denied permission after a nearly two-year wait. The only explanation he got was a letter of two sentences attached to his proposal: "Dear Mr. Foose, Please refer to the communications regarding your proposed research in Kazaringa Sanctuary. I have to regretfully inform you that it has been decided not to accede to your request. With best wishes, yours sincerely, Ranjitsinh."[10] When the Smithsonian requested permission to implement their U.S.-approved and funded plan for three large ecological research centers in India, one each in the Gir Forest Sanctuary, Corbett National Park, and Kazaringa Sanctuary, they were denied permission.[11] When the Smithsonian expressed interest in working with the BNHS in southern India on a tropical forest project, "Rajitsingh [*sic*] said that his own people and the state governments were about to undertake an extensive program there and would not appreciate any interference."[12]

Figure 10. At the edge of Kalakad-Mundantharai Tiger Reserve, a village boy stops to look at the new sign going up at the park entrance.

Cassandra Kasun Lewis, May 1998

Project Tiger

The most striking example of the move to assert IFS and Indian governmental control of ecological studies and to minimize U.S. researchers and their funding, as well as non-IFS ecologists, was the decision to allow only researchers from the Indian Forest Service to do research for Project Tiger,[13] a globally renowned attempt, started in 1973, to save the Indian tiger through a series of reserves. Its start coincided with international agreements such as the Convention on International Trade in Endangered Species of Wild Fauna and Flora, which for the first time restricted the international trade in products such as tiger pelts. Project Tiger was a landmark attempt in species conservation and brought a huge amount of international attention and money to the government of India. The wildlife department was determined that this money and attention be directed to Indian scientists. Kailash Sankhala, the first director of Project Tiger, was even intent on making sure that the credit for its inception was not given to foreigners.[14] The Smithsonian offered to send a U.S. team to help the IFS conduct a tiger census, following the 1969 IUCN meeting in New Delhi, where a tiger conservation project was formally proposed. Sankhala refused and carried out his own survey, as Ripley believed, "because of an evolving nationalist pride in the project and reluctance to accept outside help."[15] It appears that India did not want to replicate Dan Janzen's Guanacaste National Park in

Costa Rica. In fairness, although he was a forester, Sankhala was quite critical of many of his colleagues in the IFS, particularly for their neglect of wildlife concerns.[16]

In March 1972, John Seidensticker, a research scientist from the University of Idaho who had made his reputation on mountain lions in the United States, traveled to India to investigate the feasibility of a Smithsonian supported research project on tigers in Corbett National Park. The project was to use PL 480 funds, and Ripley (and the other Smithsonian scientists) hoped that the government of India would relax its freeze for such a crucial project. Ripley wrote to the U.S. ambassador to India, "I believe it is one of the few overtures toward the Indian government which the U.S. can make at this time which has international appeal, and towards which the Indian government themselves [he means Indira Gandhi] has demonstrated enthusiasm at the highest level."[17] While in India, Seidensticker met with Karan Singh, chairman of the Indian Board for Wildlife (IBWL). When Seidensticker presented his proposal and rationale, Ranjitsinh, also present at the meeting, presented a counterargument against it.[18] Shortly after his return from India, Seidensticker attended the 1972 World Conference on National Parks. He saw Ranjitsinh at this meeting, who "made it clear that PL 480 funds were out." He saw Sankhala at the same meeting, and Sankhala "was nice but made it clear he did not want Corbett 'turned into a Smithsonian laboratory.'"[19]

There were scientists and conservationists in India who did not hold these beliefs, chief among them the scientists not connected with either the IFS or the government—scientists, for instance, at the BNHS or connected to WWF-India. These Indians were not in a position to influence policy in the least, however, given the political climate of the early seventies. Not a single member of the Indian trustees for the World Wildlife Fund (which had been so essential in the formation and funding of Project Tiger) was invited to Corbett National Park for the inauguration ceremony for Project Tiger. One of these trustees (a man who was not a forester) wrote to the chairman of the Project Tiger steering committee,

> While I know that you are too broad minded to harbor any xenophobic feelings and realist enough to know that we have not the biological/ecological expertise to spurn offers of foreign technical assistance in this area, there are many in the Forest Service in Delhi and in the States whose "nationalism" appears to be clouding their judgment. . . . Seidensticker who wanted to work on the Home Range of the tiger with telemetry was cold

shouldered away and is now going to work in Nepal. Schaller is working in Pakistan though he was so keen to be given an opportunity to work in India. I think it is our serious duty as conservationists to prevent politics from invading environmental projects and to ensure that scientists of standing recommended by IUCN/WWF are welcomed in India. . . . We have now quite a remarkable opportunity to harness the best talent of the world in the sphere of ecological studies, and we must not throw it away.[20]

The U.S. scientists did not abandon their plans; they were merely diverted to other field sites.

Since India was not available, Seidensticker and the Smithsonian turned to plan B: Nepal. At the same 1972 conference , Seidensticker met and talked with Kirti Tamang, a Nepali grad student at Michigan State who was looking for a Ph.D. project. They decided that working on tigers in Royal Chitwan National Park in Nepal would be just the thing. The Smithsonian officials agreed. Ripley believed that a Nepalese tiger project, up and running, would encourage India to open up its own Project Tiger to outside scientists. In a letter to Guy Mountfort, an international board member for WWF-International who led the WWF's tiger programs, Ripley wrote:

the Indian side of the project is likely to dillydally along, and the Indian Government officials are not likely to be noted for their speed. Therefore, if we could start a project in nearby Nepal, it seems to me that it would be one of the most effective ways of showing that the World Wildlife Fund really means business. . . . Our proposal to get Seidensticker and Tamang into the field in Nepal was to be a first step to get these men eventually into action in India where Seidensticker is presently unable to operate, lacking PL 480 approval. . . . [W]e cannot just leave the project by giving the money to one or other of these Governments and go away and hope that the Tiger will be saved. We must, therefore, work out some technique by which we can encourage the possibility of foreign scientists being involved.[21]

In Nepal the Smithsonian could not rely on PL 480 funding, so they had to use funds raised by WWF-US that were originally going to India for Project Tiger.[22] In order to get foreign funding accepted in Nepal, Challinor, the assistant secretary for science at the Smithsonian, oversaw the forming of an international foundation, the King Mahendra Trust for Nature Conservation, which still exists and funds wildlife studies in Nepal today. One unintentional side effect of turning to Nepal when kept out of Project Tiger in India was making Royal Chitwan National park in Nepal (on the border with India and preserving a similar ecosystem with similar species as many northern Indian national parks) the center for almost all American and international tiger research, funded by thousands of dollars, and with access to the most up-to-date equipment available. The success of the tiger studies led to further research in this park, on rhinos and crocodilians. Chitwan is now one of Asia's showcase parks and one of the continent's most thoroughly studied ecosystems. In contrast, not one bit of published research was conducted on Indian tigers for the first decade of Project Tiger's existence.

The Indian government did not bow to continued pressure from the Smithsonian, WWF, and BNHS to allow U.S. scientists in to study the tiger. When Mel Sunquist wanted to come and do work on the tiger in 1974 with funding from the Smithsonian and the encouragement of George Schaller, he was also forced to go to Nepal and join Seidensticker and Tamang.[23] Relations between the Indian Project Tiger and the Nepali tiger projects were tense. In 1974, Sankhala asked the Nepalese government for permission to come observe the Smithsonian tiger project. The Nepalese Department of Forests would grant permission only if the Indians would allow an interchange with India's Project Tiger. Sankhala would not allow it, and he was barred from Chitwan. When Sankhala asked the Smithsonian for copies of research reports, the Smithsonian scientist with whom he spoke "indicated to him that the reports he referred to were property of the Department of Forests, HMG Nepal." The Smithsonian might have even felt a little smug: "Sankhala said he was slightly dismayed at the progress of Project Tiger in that administrative details had tied the project up for almost a year and only this month were they able to formulate some management plans."[24] Then, in August 1974, it seemed that the U.S. tiger scientists got their first opening.

In the swampy Sunderbuns sanctuary of India, near the border of Bangladesh, a tiger living near a village had attacked and killed a woman. State forestry officials decided that the best plan was to tranquilize the tiger with a dart and relocate it to an area far from human intrusion. The necessary equipment and expertise were just across the border, in Nepal. There was a flurry of telegrams back and forth between India, Nepal, and the United States, gaining the necessary permissions. Mountains of red tape were miraculously dissolved and within a short time John Seidensticker found himself sitting in a blind at night, waiting to dart this killer.

The Indian press went wild, covering the affair in great detail and describing Seidensticker as a latter-day Wild West cowboy. After a few nights he darted the tiger and it was relocated. Although a few discouraging words were heard (particularly about Seidensticker's failure to dart the tiger right away and his initial mistaken estimate that the tiger was a female—it was a young male) the operation was judged a success. Then, on August 28, a few days after its release in its new home, the state forestry officials and Seidensticker found the tiger dead.[25]

The press turned against Seidensticker with fury. Seidensticker and the forestry officials determined that the young male had been killed by a larger male, who had perhaps been adverse to sharing its territory. Newspaper articles attempted to link Seidensticker with the CIA. Critiques of Project Tiger were rampant. And Sankhala declared that the cause of death was certainly the tranquilizer, not another tiger: "Tigers are non-territorial. They do not resent intrusion or jealously guard their territory."[26] Sankhala suggested that the tiger had died because its body had not been able to recover from the shock of drugging—perhaps the dose was too high?—and that the marks on the body that Seidensticker had seen were made by wild boars.[27] An overaggressive and incompetent U.S. expert had killed the tiger, not nature. Seidensticker quickly left India—the U.S. embassy in New Delhi recommended that his stay in Delhi need not exceed twenty-four hours.[28] Any hope for a tiger détente was shot.

The Indian Forest Service: Managers for India's Protected Areas and Species

Enhanced awareness of conservation concerns in India in the sixties and seventies and the formation of Project Tiger were mirrored by an increased concern within the IFS for managing India's protected areas. In the Indian central administration, the IFS has complete responsibility for all reserved forests, national parks, and sanctuaries, as well as the protection of endangered species. The IFS combines what in the United States would be the Forestry Service, the Fish and Wildlife Service, the Bureau of Land Management, the National Park Service, and every state's individual department of natural resources. The IFS is deeply involved with contracts for lumber extraction, and it controls access to minor forest products of vital interest to tribal and peasant populations. The IFS is a very large, very diverse, and very complicated bureaucracy.

Members of the IFS are selected on the basis of a highly competitive national examination, but they need not have a background in forestry. Upon selection, new members of the IFS are trained at the Forest Research Institute, which, like the WII, is located on a campus outside Dehradun. When they complete their training, they are assigned to the forestry department in one of India's twenty-six states. Although the IFS is a federal (national) administrative service, individual foresters are assigned to specific states for their careers and will always work for that state's wildlife service. (IFS officers may be deputed to the central headquarters in New Delhi.) Within their given states the forest officers are assigned to specific posts that they typically hold for three to five years before being rotated to another.

The Indian Forest Service, which is a continuation of an organization established by the British in the nineteenth century, has its own tradition of naturalists apart from the BNHS and CES. Many foresters, working in reserved forests for their entire careers, develop an impressive understanding of India's flora and fauna.[29] There is a research branch of the IFS, but it is almost exclusively devoted to silviculture (the cultivation of forests). There was also a game orientation among IFS foresters, an interest that passed directly to the IFS after independence. As late as the mid-sixties, it was considered almost mandatory that young forest officers "bag" a tiger within a few months of their first posting.[30]

This orientation resulted in the development of expertise in managing some big-game species. In a time when few Indians were actively pursuing ecological research, foresters were often the de facto experts on Indian ecology. This expertise was usually practical and oriented toward goals more derived from tradition than conservation science.[31] By the mid-sixties, some members of the IFS wanted to develop their expertise in conservation management and science further.

Institutionally this growing interest in conservation was indicated by new wildlife management courses at the Forest Research Institute (FRI). Such courses had been introduced in 1955, and in 1970 a certificate course in wildlife management (six months long) was begun. When S. R. Chowdhury, the man who had started this program, left in 1974 he was succeeded by N. R. Nair, who was named the first director of wildlife research and education. Nair began a ten-and-a-half-month postgraduate course for foresters specializing in wildlife management. Tragically, he was killed by an elephant while on a trip with his first group of trainees. He was succeeded by V. B. Saharia, who served as director until 1982, when his title was changed to the director of the newly formed Wildlife Institute of India. The

WII was born out of the IFS training program at the FRI. Saharia served in this post until the institution was granted autonomy in 1986.[32]

The 1974 transition to an extensive postgraduate (i.e., after completing the normal forestry training school) program in wildlife management occurred shortly after the passage of the 1972 Wildlife (Protection) Act and the beginning of Project Tiger in late 1973. The act included a list of endangered and threatened flora and fauna in India, and it codified into law the criteria for establishing and managing a protected area such as a national park or wildlife sanctuary. According to the act the management of all protected areas was to be the sole responsibility of the IFS; the conservation of endangered species was also placed within the forest service's domain. When Project Tiger began, the IFS again was given complete responsibility for its implementation and the new tiger reserves were simultaneously designated as national parks or wildlife sanctuaries. All Project Tiger research staff positions were reserved for IFS officers. This specification, in conjunction with the restriction on U.S. scientists at that time, meant that in 1974 there was a sudden and urgent need for forestry officers trained in wildlife management and research to implement Project Tiger and to manage India's other protected areas in accordance with the act.[33]

Such expertise was slow in coming. The first certificate course, Nair's only one, was offered in 1974–75, and it is not clear if it was ever concluded. With Nair's death, there was an understandable break in the running of the certificate program. The second program did not take place until 1979–80. These courses did not train many students (only 47 between 1974 and 1982), and graduates were very junior when they left for wildlife management assignments.[34] Many of the research posts in Project Tiger reserves were left unfilled. Of those that were filled, many were occupied by foresters whose attention was diverted by having to assist in the running of the reserve and by conducting annual tiger counts. The counts were highly controversial and nine different interview subjects, including biologists and forest officers, told me (off the record) that there was pressure to show a steady improvement in tiger populations every year. Valmik Thapar, a member of the Project Tiger steering committee and producer of the BBC documentary *Land of the Tiger* (1998), discussing corruption in the state of Madhya Pradesh, charged that, "the government 'increased' the number of tigers in the 1997 census by 10, just to show progress."[35] Beyond counting the tigers, most research officers conducted no formal research.[36] Despite ten years of Project Tiger, even very basic information about tigers in India remained unknown, as it had been before 1973.

The one exception was Ullas Karanth, who has been allowed to conduct research on tigers in Nagerhole National Park since the late-eighties (the first person allowed to do so who was not in the IFS). Karanth, who now serves on the Project Tiger board, agrees that "the scientific program has been bad. . . . It is very difficult to combine [forestry and enforcement] with creative science and thinking."[37] This reality, added to the isolation in which most foresters work, has made it nearly impossible for foresters to conduct research projects. Even those forest officers who are interested in research and have an idea of how to do it are foiled by their heavy management workload and by the rotational system of the IFS, which discourages long-term stays in any given park. The IFS by its practices has thus created many knowledgeable naturalists, but no true research ecologists.

Founding the Wildlife Institute of India

V. B. Saharia, director of the wildlife research and education program of the IFS when it became the WII, was one of those naturalists, and H. S. Panwar, who followed Saharia as director of the WII in 1986, was another. Panwar had been the IFS officer in charge of Kanha National Park (where Schaller researched *The Deer and the Tiger*) in the late sixties and throughout the seventies. Although Panwar's tenure at Kanha did not overlap with Schaller's stay, Panwar was at Kanha when Claude Martin, now the director of the WWF-International, based in Switzerland, did a study for his Ph.D. on the barasingha (swamp deer).[38] Panwar oversaw the management of Kanha during the period when it was declared a tiger reserve, and it was on his watch that the last villagers were removed from the park. At that time Panwar became interested in tiger ecology, and he claims that while at the park, "I was pursuing my own research, conservation biology, especially tiger ecology, and that put me in touch with the scientific community."[39] Panwar also hosted the trainees from the Forest Research Institute's wildlife management certificate course when they came to Kanha on field trips. Kanha is one of India's flagship national parks, and Panwar was so successful at this post that he was allowed by IFS leadership to stay there for thirteen years. In 1981, Panwar was moved to Delhi and became the director of Project Tiger until he left to head the WII in 1986.

In Delhi, Panwar met a British scientist, John B. Sale, who worked for the Food and Agricultural Organization (FAO).[40] Sale had been assigned by the FAO to work on another set of highly endangered animals. Project Crocodile began in 1974, right after Project Tiger, but with considerably less fanfare or governmental attention. Also unlike Project Tiger, the emphasis in Project Crocodile, supported in part by the FAO, went beyond protecting natural environments and included research on captive breeding and rerelease into the wild.[41] In addition to Sale's participation in the project, a number of graduate students were recruited to do research on the three crocodile species found in India. The Central Crocodile Breeding and Management Training Institute was established in Hyderabad, in southern India, and the researchers worked out of this office.

Project Crocodile was an unqualified success. Eventually, the problem being investigated by crocodile researchers was not how to save the species, but how to get rid of them.[42] Beyond captive breeding, the researchers gathered very thorough information on crocodile life histories. These graduate student researchers led the first large-scale species study conducted by Indians in India, starting in 1975. By the time they had finished their research, in 1980, there was no danger that any of the Indian crocodile species would ever become extinct, though their continued wild existence is still in question. The IFS participated in Project Crocodile, and many foresters went to the Central Institute for training, but the forestry department did not control the research aspect of the program.[43] The contrast with Project Tiger is stark—Project Crocodile included international funding and a foreign scientist to help train the graduate students and oversee the project, it was research driven, and it succeeded. This is not quite a fair comparison, however, as crocodile conservation, internationally, has been much more successful than mammal conservation, in large part due to the greater ease of raising crocodiles in captivity (they lay lots of eggs!).

When Panwar and Sale met and talked about their respective projects in Delhi in 1981 and 1982 (Panwar as director of Project Tiger, Sale as chief technical advisor for Project Crocodile), they agreed that there was a need in India for an institution that could both train forest officers in wildlife management and conduct research (geared toward management) on India's wildlife. Panwar remembers, "People were talking about it . . . and since I was in the ministry I was able to pursue it." At the same time naturalists in Delhi were planning a special Indo-U.S. workshop on the techniques of wildlife management and research. Sálim Ali had begun suggesting in the late seventies the need for a workshop in which top U.S. ecologists could demonstrate current field techniques to Indian colleagues.

By 1979 the freeze on U.S. scientists and funding in India was lifting. Sarah Blaffer Hrdy received permission to work on langurs (monkeys)—the first Smithsonian project approved since the early seventies.[44] The Smithsonian was a minor player now, however. The government of India had decided to allow the U.S. Fish and Wildlife Service (USFWS) to be the primary collaborating institution for ecological research in India. I did not find direct evidence related to why the government of India made this shift, but some contemporaries at that time thought that the government preferred to collaborate directly with a U.S. governmental agency, who's funding and management were a matter of public record and under political control. In November 1979 the USFWS signed a contract with the BNHS allocating over $340,000 to the BNHS bird-banding study, through the PL 480 funds. Ali was in shock by the size of the grant.[45] The government of India had decided to spend the PL 480 funds down as rapidly as possible, and they approved a number of grandiose projects in the late seventies and early eighties toward this end.

In this thawing climate, Ali's suggestion of an Indo-U.S. scientific conference "received the support of [newly returned to office] Prime Minister Mrs. Indira Gandhi who headed the Indian Board for Wildlife, the apex advisory board in this field in India." The Indo-U.S. Sub-Commission on Agriculture gave its support to the idea at its September 1980 meeting in Delhi, at which the participants "strongly confirmed the relevance and importance of such a workshop." Steve Berwick, a U.S. ecologist who had worked in India on ungulates and vegetation surveys in Gir National Park, and V. B. Saharia, at that time director of wildlife research and education at the FRI, were selected as the cochairs for this workshop, which was to be held over three weeks in January 1982. They selected "fifteen top U.S. wildlife scientists and fifteen specially identified Indian wildlife managers and scientists" to present papers at the workshop, and decided to hold the workshop at Kanha National Park.[46]

This workshop was well attended by "nearly 100" participants, from India, Nepal, Bangladesh, and Sri Lanka. The workshop had two goals: to facilitate "modern technology transfer from the U.S. scientists" and "to develop . . . perspectives for a scientific approach to wildlife management, related to applied and basic research and suited to the conditions of this region."[47] Thus, after nearly a decade of limiting the role of U.S. ecologists in India, the formal reintroduction of U.S. scientists and techniques into India was announced—a reintroduction originally suggested by Sálim Ali. Unlike earlier exchanges, though, this workshop was highly formal and explicitly demanded equal participation; further, a conscious goal was the adaptation—not blind acceptance—of U.S. techniques to Indian conditions. Over a decade later, the workshop proceedings were adapted into a manual, *The Development of International Principles and Practices of Wildlife*

Research and Management: Asian and American Approaches (1995) which is still used as a text for WII courses. As Saharia mentions in his recounting of the history of that workshop, the original title of the manual was to be *Manual of Techniques in Wildlife Research and Management.* The change of title indicated "that the requirements of wildlife and wild land work, outside Europe and North America are identifiably different and dictate different approaches."[48] What this means is that U.S. techniques were not to be imperialistically imposed on Indian scientists and managers, they were adapted to local needs.

A primary recommendation of the 1982 workshop was "the establishment of a full fledged Wildlife Institute of India." Since administrators had been talking about this for over a year, the high-profile workshop recommendation was enough to push the idea on through. In May 1982 (just five months after the workshop), the government and the IFS agreed to merge the Directorate of Wildlife Research and Education at the FRI in Dehradun and the Hyderabad Central Crocodile Breeding and Management Training Institute. This was now the WII. Saharia was named director of the new institute, which primarily served as a training site for foresters. The Hyderabad institute was kept open for four more years as a southern training site, no longer focusing just on crocodiles but on protected area management. Eventually, the Hyderabad staff moved north and joined the Dehradun-based arm of the WII. The WII continued to act as a subsidiary of the FRI until it was granted autonomy in 1986, but it was not until 1992 that the WII moved to its own campus. As an autonomous educational institution of the government of India, the WII was authorized to develop as a research institution, to supervise graduate degrees (M.Sc., Ph.D.), and to accept grants and funding directly.[49]

A. J. T. Johnsingh: Biologists Enter the WII

After gaining autonomy, the WII still had a strong forestry emphasis. It continued to offer training courses for foresters.[50] The director, Panwar, was a senior forester without graduate training in biology but with many years of experience in Kanha National Park and as head of Project Tiger. Many of the faculty members were foresters on deputation from the IFS for a three- or four-year term; most lacked training in biology. Thus when Panwar pushed to establish a faculty divided into three groups— biology, management, and extension—it was the foresters who dominated the management and extension faculty. The biology faculty had to

be recruited. The biology faculty, who would assume the primary in-
structional duties for the graduate students and conduct research, were to
be formally trained as biologists, not foresters with added qualifications.
The first to be hired, in March 1985, was A. J. T. Johnsingh, who became
the head of the biology faculty.

Johnsingh's path to leadership at India's new premier wildlife biology
institution in 1985 was indirect and relied on the ecological expertise that
had developed within the BNHS and CES in the sixties and seventies. In
May 1971, Johnsingh was on vacation from his post as a zoology instruc-
tor at a southern Indian college. He had been raised in a family that had
always enjoyed "outdoor" life, including fishing, hunting, and camping in
the forests, so he decided to spend his vacation in the Western Ghats—a
southern range of mountains running parallel to India's western coast.
Johnsingh remembers walking through the rain forest, where he "met Mr.
J. C. Daniel; we were at an altitude of about three thousand feet." Daniel
was Sálim Ali's able lieutenant within the BNHS, and has served at various times through the sixties, seventies, eighties, and nineties as its curator
and honorary secretary. In the course of their conversation, Daniel asked
Johnsingh about his background, "So I told him that I had done my M.Sc.
in zoology and was teaching in a college. Then Daniel said to me, 'Why
don't you take up some studies in wildlife?'" Johnsingh replied that, although he was interested in wildlife, he had no formal training.[51]

In 1972, nearly a year later, Johnsingh received a letter from Daniel.
An expert on canid behavior, Michael Fox, was coming to India from the
United States to study the dhole, a wild dog native to India. Fox had been
trying to get into India since 1969 but could not get access to rupee
funds—he finally decided, "As things stand now, I have no financial support whatsoever, but since the survival of the Dhole is in jeopardy, I feel
that every effort should be made to get some field work accomplished."[52]
As he was only affiliated with his home university (not the U.S. government or the Smithsonian), did not require any rupee funds, and was staying only for three months, he was allowed to enter India. Fox was quite
unusual both in being willing and able to pay for his research out of his
own pocket and in being given permission to do work in India in 1973
and again in 1975. He illustrates that not all American scientists were kept
out of India's forests, just the vast majority.

Fox was affiliated within India with the BNHS, and as per the Indian
government's guidelines, he needed an Indian research assistant (collaborator) for his work: "It was almost mandatory that when a foreigner comes

and works here, our government expects that he should provide technical assistance and experience and some funding also for an Indian assistant." Because Fox intended to work in Mudumalai National Park, in the state of Tamil Nadu, it would be helpful if his Indian research assistant could speak Tamil. So, Daniel asked Johnsingh, "Would you like to join him?" Johnsingh agreed to help, if Daniel could persuade the college to release him from his teaching duties. Daniel did so and for three months in 1973 Johnsingh took an unpaid leave to assist Fox. Two years later, in 1975, Fox returned to India for a slightly longer stay, and again Johnsingh was his assistant.[53] This time, though, his college told him, "This is the last time we are giving you leave like this. And if you want to do anything more like that, you will have to cut off your association with the college."

When Fox's research ended, Daniel encouraged Johnsingh to pursue a Ph.D. in wildlife ecology. Johnsingh replied, again, that he was interested, but that he had no funding, he would lose his job if he did not return, and he had no training in wildlife research other than what he had learned from Fox. Daniel was unswayed, and "he helped me prepare a proposal which solicited funding from WWF-India. And at that time, the WWF-India office was in the BNHS office, at the Hornbill House. I received funding." Johnsingh spent the next two years (1976–78) conducting research on the dhole for a Ph.D. from Madurai University.[54] Johnsingh is the only ecologist I met who had actually received his start as a result of the Indian government's requirement that U.S. ecologists work with an Indian field assistant. Johnsingh's success is less a testament to the government's policy than to his own determination. Johnsingh claimed, "No one grew up; the people who worked with these foreigners, unfortunately they did not carry on. Those of us who did later work, did it through our own efforts," not with any continued assistance from the government or from their former U.S. colleagues. Still, it seems clear that Johnsingh personally was given at least an initial push into wildlife biology by his association with Fox.

The Indian educational system does not require any coursework for a Ph.D.; students are simply asked to do a research project and write a dissertation. Thus, Johnsingh took no courses in field ecology or wildlife biology. After obtaining funding from WWF, he enrolled in Madurai University and began his fieldwork right away. While Johnsingh had been able to learn some field techniques from Fox, he was mostly on his own. With this lack of formal supervision and training, Johnsingh relied informally on Madhav Gadgil at the nearby IISc (this was before the CES had been formed) to advise him and assist him in writing his thesis. Johnsingh even stayed at the IISc for a period after his fieldwork was done. Gadgil was particularly useful in helping Johnsingh become familiar with the techniques and current literature of field ecology. Johnsingh was the first Indian to complete a Ph.D. in India on a wild, free-ranging mammal.[55]

During his fieldwork on the dhole, Johnsingh had met two visiting U.S. ecologists, Eisenberg and Kleinman. As Johnsingh claims, in his typical self-deprecating fashion, "They saw me and thought that I needed some [additional] training. So they asked me whether I could go to the Smithsonian." Following the completion of his Ph.D., Johnsingh received a postdoctoral fellowship to work at the Smithsonian research station at Front Royal, Virginia. Over the course of eighteen months Johnsingh worked on raccoons and opossums: "they were teaching me methods to trap them, radio-collar them, and use radio telemetry." When he returned to India in 1981, Johnsingh was for a brief period unemployed. He had given up his teaching job to pursue research and there was nothing immediate to take its place. In 1982 he obtained a research position with the BNHS, working for the next three years on the Asian Elephant Project. (He was unable to use any of his new training in radio telemetry at this time, for the project was not funded well enough to purchase the equipment.) The BNHS Asian Elephant Project was a ten-year project funded by the USFWS that called for fieldworkers throughout India to survey the status of India's elephants (receiving another of the huge PL 480 USFWS grants). Researchers like Johnsingh were hired, and then when they got a better job offer (as with Johnsingh's at the WII) they were happily bid farewell. As it had done for so many unemployed Indian ecologists through the years, the BNHS helped fill the gaps by offering them fieldwork jobs funded by international grants.

Johnsingh was hired by the WII in 1985 at the beginning of a three-year push that saw the faculty fleshed out to include seventeen members by 1988. Among others, Johnsingh's colleague on the BNHS Asian Elephant Project, S. Chowdhury, was also hired to join the WII management faculty (because he had received his Ph.D. while working with forestry officers on Project Crocodile, they placed him in that division). There were also three FAO-supported outside experts on staff with the WII (one for each faculty division, including J. B. Sale in biology)—between 1983 and 1988 the FAO-UNDP Assistance Project for the Establishment of the WII gave the institute $1,879,229.[56] That paid for the salaries of the FAO experts (one Briton, one American, and one Anglo-African from Tanzania, W. A. Rogers, who became a crucial guiding force in the early years of the

WII), as well as for computers, books, journal subscriptions, lab supplies, jeeps, field equipment, and for the travel costs of WII staff who went abroad for training.[57] In August 1985, Ph.D. researchers had begun interviewing to join the WII, and the WII opened its doors to students in the master's program in wildlife biology in January 1988.

While the FAO-UNDP project was funding institution-building measures, the USFWS began to fund collaborative research projects for Ph.D. students. When the FAO-UNDP grant ended in 1988,[58] it was replaced by an Indo-U.S. faculty development project that funded U.S. experts (not graduate students anymore, but full faculty members) coming to India. Typically they would advise on a research project, then hold a workshop, and then invite their Indian collaborator to the United States.[59] The USFWS and U.S. Forest Service continue to fund Ph.D. research at the WII to this day, and each U.S.-funded research project must have a U.S. collaborator. Following the promise of the 1982 workshop at Kanha National Park, U.S. funding insured that American ecologists have seen their access to India's national parks and wildlife restored, albeit in a more supervisory and collaborative role—rather than U.S. graduate students cutting their teeth in the fields of India, it is now U.S. professors helping to advise Indian graduate students working in the field.

Nearly all the young Indian doctoral researchers who came to the WII in the initial years worked with Johnsingh. Unlike the WII management faculty, who were occupied with running the certificate courses for the foresters, the biology faculty were quite small and oriented to research. Johnsingh was the only one in the biology faculty who had experience with the animal studies that dominated doctoral research projects. Beyond that, he exerted a certain charismatic appeal over the young ecologists. Almost every current student at WII views him as a walking legend and a naturalist without peer. One of Johnsingh's former students (who has since been hired as a WII faculty member, as have three others on the ten-member biology faculty), Ravi Chellam, explained, "He is our first trained Indian wildlife biologist. Sálim Ali and other renowned Indian naturalists preceded him, but nobody was really trained. If they had degrees, they were all honorary degrees, rather than a proper peer review process."[60]

The "explosion of people" that Chellam refers to in the epigraph to this chapter was one he helped begin in 1985: aspiring biologists from across India came to Dehradun and the WII; after training they then scattered across the length and breadth of the country to conduct research in

India's national parks and sanctuaries on endangered and threatened species. Within the first year of the WII's autonomy its researchers began research on large mammals, ungulates, and local people in Rajaji National Park; on grizzled giant squirrels in Tamil Nadu; on the behavior of the nilgiri langur in Mundantharai Tiger Reserve, Tamil Nadu; on the flying fox (fruit bat) near Dehradun; on the snow leopard populations in Ladakh; on human-lion conflicts in the Gir National Park; on a rhino reintroduction program in Dudhwa National Park; and on a biogeographic strategy for endemic plant conservation in the eastern and western Himalayas.[61]

One of the explicit goals of the Wildlife Institute of India was to conduct and coordinate applied wildlife research on Indian wildlife. This has often meant, and still means, conducting the first-ever study of an Indian animal or conducting the first study of that animal since a U.S. or European study in the sixties (and invariably looking at a different element of its behavior or ecology). S. P. Goyal, the third biology faculty member hired by WII, reviewed the state of wildlife biology in India:

> I personally feel that, except for the past five to eight years [since the WII was founded], there was not any good recent work in India. When we used to teach the M.Sc. students, we would rely upon Schaller's book *The Deer and the Tiger,* and if you wanted to look at some of the species, like . . . the blackbuck, there is not any such good work, up to the last ten years. We used to rely exclusively upon the work done by these Texas A&M biologists; for instance, they did a big monograph on nilgai and some of the other ungulates. So we had to depend much more upon the . . . Americans' work. There is a lot of it. And ecological theories have been tested in the wild. Nowadays we have a lot of information, though, and we try to supplement our teaching with examples from the Indian context. But if you are talking in 1986, we had to collate or collect our information from books like that. . . . Talking about ecological succession in animals, we used to take some of the examples that had come from Africa, like in the Serengeti, because what we wanted to explain was the principle—what is a grazing succession in animals like? So we had to bring the actual data from there.[62]

The Texas A&M ungulate biologists, incidentally, had also asked to come to India in the early seventies and been denied—they thus did all their work on Indian species that they introduced on Texas pastures and observed there.[63]

Johnsingh agreed with Goyal's assessment. When asked to characterize the differences between wildlife ecology as practiced in the United States and India, he replied that for

> any species which we look at, there is a lack of information. Anything you
> do, you will be the first to do it. So we are spread out: . . . do a two-year
> study, a three-year study, like that. But [in the United States] it is much
> more focused. The white-tailed deer, for example, there have been a thousand studies of it. And the grizzlies or wolves in Alaska. They have been
> doing that for fifty years now—in 1944, the first book about wolves in
> McKinley National Park, now called Denali National Park, was written by
> [D. H.] Pimlott. And when you look at the resources, the trained personnel, the [relatively fewer] species, they could focus and do more in-depth
> studies on cotton-tailed rabbits or big-horn sheep.

Johnsingh believes that Indian wildlife biologists can import techniques from the United States, such as radio telemetry, but as long as so much natural history information is still unknown for Indian flora and fauna, Indian wildlife biology will of necessity be oriented toward collecting basic natural history information, no matter what the current U.S. trends might be. About his Smithsonian postdoc, Johnsingh commented, "the main thing is I learned about radio telemetry, and I can use it for the lion study. . . . That field knowledge which I gathered there, it is very important now. . . . And I also could bring a lot of literature back which is still being used by the students."

Conceiving a Better Indian Wildlife Biology

The biologists who work at the WII, most notably Johnsingh, have played an important role in converting it from a strictly forestry training program to one known (particularly internationally) as a wildlife research institution. They have begun to realize the original idea of IFS foresters, naturalists like Saharia and Panwar, who saw the need for increased knowledge about India's wildlife and different ecosystems in order to effectively manage India's protected areas. Yet one previous administrator from the WII described how research was valued only insofar as it contributed information to teaching and to management:

> We needed training institutions because we have four to five hundred-
> odd protected areas, even if we consider [only] two hundred of them as

important from the active management angle. And then we needed maybe a couple of thousand people to manage these protected areas. So we needed a training institution which will give . . . the trainees the skills to manage these areas. . . . and we always thought that teaching has to be contemporarily relevant both in terms of the advances in science as well as the changes in the ground situation. So, suppose if we had faculty who concentrate only on taking these trainees out on tour, bringing them back, and just teaching them? After some time, the faculty for the institute will probably fall out of step with what is happening. So we at the same time mounted a research program. We thought that, well, we are talking of faculty development, we are talking of at least 30 percent of the time devoted to research, which is management research, substantially, but not totally. Basic ecological, basic biological, research is also done. But a lot of conservation biology research, and similarly a lot of park management, habitat management, that kind of research. . . . So these people are always abreast of the ground realities; they are able to lace their teaching with examples from the field with which they can go and tell people what is happening.[64]

It is possible that this view is partially motivated by a belief that the IFS will ultimately be able to conduct much of the important research on India's wildlife through its very proximity to the field. Another former WII administrator is reputed to have said to a young biologist, "whatever you do in four, five, or ten years on the tiger, I can write in thirty minutes; working on tigers is not in the WII mandate."[65] That administrator was insisting that his years in the field as an IFS officer, observing tigers, had been the equivalent to formal biological research, which was thus not needed. He and the other foresters involved with Project Tiger had things covered. It seems likely that many foresters working at the WII still held out hope that the IFS would be able to assume a significant research role with the help of increased training of officers at the WII and perhaps did not want nonforestry biologists to fill that niche. In fairness, though, the first administrator quoted above was quite proud of the WII researchers who had since become biology faculty members of WII.

The first annual report of the newly autonomous WII in 1986–87 was less direct about the role of research in facilitating teaching, though it did claim, "Research activities . . . are also looked upon as a means of faculty development, essential to this upcoming Institute and to the upcoming Indian wildlife science which it is required to foster." The report also specified that research was to be applied rather than basic research: "wildlife research must address endangered ecosystems and species, technique development and application of modern technology under Indian conditions and, importantly, the human dimensions aspect of wildlife conservation and management." This view followed directly upon the recommendations of the 1982 Kanha workshop. The research of WII was designed to help manage India's protected species and parks. Johnsingh fundamentally agrees with those aims: "This is a very important institute. We train the people who are the wildlife managers, and conservation cannot be done only with writing papers and classroom lectures. It has to be practiced in the field. And who is going to practice it? Only the forest officers who are working in the protected areas as managers. . . . So I thought working in this institute is an excellent opportunity to train people. We have the direct application of conservation." It was not the opportunity to do research that drew Johnsingh to the WII as much as it was the opportunity to train the people who would manage the natural areas of India, which he so loved. Johnsingh is realistic about who holds the power to either manage effectively or destroy India's protected areas, and he wants to have the chance to train—to "convert"—them.[66]

In the nineties the younger WII biology faculty (most trained by Johnsingh at the WII) have developed their own agendas. The students of the old mess called these people the young turks (which amused two Turkish women taking classes with the M.Sc. students). The reference group of the younger faculty were not foresters or policymakers but other biologists, not just at the WII or in India, but throughout the world. With funding from the USFWS, USFS, and the FAO-UNDP project, almost every one of the biology faculty have spent time overseas, and all have collaborated with American and European scientists.

The WII has thus become an institution with two cultures: one based on the forestry training courses, largely staffed by current foresters on deputation or by ex-foresters, and the other comprising the students in the M.Sc. program in wildlife biology and the Ph.D. researchers, trained by Johnsingh and his colleagues in the biology faculty. Insofar as there were conflicts in the sixties and seventies between the IFS and Indian ecologists outside the IFS, who worked in IFS-managed sanctuaries and made recommendations about the management of protected species and areas, those conflicts are now internalized to some degree in this one institution.

The two diploma courses for foresters are internationally recognized and have included participants from throughout Asia and Africa. The WII is still supported by the Ministry of Environment and Forests, and the leader of the WII will always be supplied from the IFS. When Panwar left to join the FAO in 1994 (he currently works as an external advisor in Sri Lanka) he was succeeded by S. K. Mukherjee, a former forester with a long background in national-level environmental planning who had worked at the WII since 1987. The research program, though, is what has given the WII most of its recognition in the West. The students and most of the biology faculty attempt to publish in the top conservation biology journals in the United States and Europe. In the present as in the past, the collaborative projects between U.S. scientists and WII scientists always include student researchers, and the student researchers work with the biology faculty. Thus most of the WII's international collaborations are in effect collaborations between members of the biology faculty and U.S. scientists. This further cements the view, outside India, of the WII as a conservation biology research institution.

In my many conversations and interviews with the biologists and students of the WII, it was obvious that they agree that scientific information is needed to manage India's protected areas and species and that much more natural history information has to be collected. They all are committed conservationists, in the sense that they value the preservation of natural ecosystems in the face of potential destruction, though they differ at times on what action is called for. They also agree that teaching is strengthened by research and that well-trained managers are needed for India's protected areas. The biologists transcended the local WII mandate, though, in their ambition to produce scientific information of international significance. Many of them have internalized the norms of the international practice of conservation biology, and many speak of their desire to improve the level of biological research conducted at the WII (by improving, they mean bringing it closer to the norm as seen in U.S. and British publications). A surprising number of the M.Sc. students finish at the WII and go to the United States or United Kingdom for their Ph.D.s. Some members of the current biology faculty expect that their research will not only improve park management but that it will contribute to global discussions of conservation biology. As just one indication of this focus, all M.Sc. students are required to submit their master's theses to a major journal before they are given their degrees. With that aim of conducting science by international standards, WII biologists are willing to look critically on their own scientific practices.

An Experimental Wildlife Biology

Different faculty members understandably have different images of what their ideal science might be. Common among some of them is the criticism that WII biology is insufficiently experimental, in comparison to an idealized U.S. ecological practice in which experiments are routinely used to gain more specific and useful information.

WII biologists are aware that experimental wildlife biology can be a useful tool for management. At Yellowstone National Park, for example, there has been a simple ongoing experiment in which elk are kept out of enclosures. Vegetation is then compared within the enclosures and without, and the data are used to see what effect elk grazing has on vegetation.[67] Simple experiments such as this give researchers and managers comparative data that would not be apparent from observations alone. One WII researcher suggested that one easy way for Indian scientists to get data of this sort would be to treat normal park management approaches in India as if they were experiments—testing the effect, for instance, of banning grazing in a national park or of building a road in a formerly pristine area.

Sometimes ecologists look for what they call natural experiments, that is, naturally occurring differences (or differences caused by historical human disturbance) that can be compared and then theorized about. One famous example of this was the island biogeography research of Wilson and MacArthur, who used data on species richness, extinction rates, and colonization rates on different islands in the Caribbean to derive their theory relating island size and number of species, and linking extinction and recolonization rates. Of course, Wilson and Simberloff then resorted to a manipulative experiment to test the theory of island biogeography when they fumigated mangrove islands in the Florida Keys.

Directly manipulative experiments are often out of the question for WII researchers. Because they so often are working on endangered species or in protected areas, the law prohibits them from making any (potentially harmful) modifications. Even radio-collaring of endangered animals is highly controversial, as with Williams's difficulty in getting permission to radio-collar elephants. Natural experiments are often the only truly viable research option for WII biologists who hope to do a more experimental research program on an endangered species, or in a protected area.

Within the M.Sc. program, the WII students are encouraged to conduct experiments for their thesis work: "If they do natural history kinds of stuff, then it would not be as much scientific training as compared to if they set up an experiment, go ahead and collect data, and then test hypotheses. So we encourage them to do that, and most of them have, this last batch, except for two or three people." Experiments seem more properly scientific than just observational ecology—they follow the scientific method, after all. Similar views are held by U.S. biologists, who encourage experimental ecology at the expense of more descriptive (even if theoretical) ecology. One U.S. ecologist who works on the fifty-hectare forest plots in Panama has been accused of doing work that is only descriptive and not experimental. He counters, "But I think the proof is now coming in that you can in fact test hypotheses extremely rigorously with these data sets. . . . When you have a total sample, nobody can say that you have the wrong answer for that sample. And so we have just overpowered the arguments by sheer force of demonstrating the utility of these data." This scientist claims that ecologists are routinely underfunded within the United States because their work is so often based on collecting observational data sets, as opposed to doing experiments and because most general-science funding boards do not recognize observation as valid science. If relatively well positioned U.S. ecologists have difficulty obtaining funding for observational ecological work, it seems obvious that this problem is even more acute in India.[68]

Late in 1990, Ravi Chellam and Ajith Kumar learned that the USFWS was prepared to fund new research projects at the WII. Kumar had recently finished his Ph.D. and Chellam had completed his research but had not yet written his dissertation. They decided to write a proposal for a four-year research project that Chellam could conduct (under Kumar). In early 1991 they submitted their proposal, in which they suggest taking advantage of a natural experiment to test island biogeography principles, as applied to reptiles, amphibians, and small mammals. By comparing reserves of various sizes (11 ha to 500 km^2 of forest) and their associated species, they could "predict the changes in faunal distribution and abundance as a result of fragmentation across the Western Ghats."[69]

In effect, Chellam and Kumar proposed a terrestrial version of MacArthur and Wilson's 1967 island study:

> I was definitely interested in the Western Ghats and rain forests. And [Kumar] was interested, because he had done his Ph.D. with the lion-tailed macaque there in the rain forests. So both of us sat and talked about these issues. And this was a thing that was then of global importance. Almost

everybody was trying to look at rain forest–related issues, and [the theme of habitat] fragmentation was definitely flowering, and its time seemed to have come. And we, in many ways, had this kind of a designed set-up in the sense that you had these natural—well, not natural—fragments which were already existing, and you had a relatively contiguous forest. So it just seemed to be too good an opportunity to miss.[70]

As Chellam and Kumar were aware, other experiments in the eighties were also testing habitat fragmentation. The most famous was Thomas Lovejoy's experiment in the Amazon basin, still ongoing, that compares the effects of habitat fragmentation on pieces of rain forest varying from tiny to several square kilometers. Lovejoy's experiment is manipulative: the government of Brazil has designated certain patches of forest to be left untouched by loggers, who otherwise practice clear-cutting. The project was also extremely high profile. It was partially funded by the Smithsonian, where, along with the WWF-USA, Lovejoy was employed during the eighties. Studies of rain forest fragmentation were hot. Chellam and Kumar hoped to draw upon this tradition in conducting a similar project in India, oriented toward relatively understudied species.

Chellam and Kumar were funded by USFWS, though it took five years for all the Indian government's paperwork to clear. Finally, when the project began, in 1996, Kumar had moved on to the Sálim Ali Centre for Ornithology and Natural History in Coimbatore and Chellam had joined the faculty at the WII. They were told that their project "was the most well-written proposal," and they have continued to shop it around to other funding agencies to raise still more money. By March 1998, they had received additional money from the Wildlife Conservation Society and Operation Eye of the Tiger (both U.S.). (The latter funded them because some of their work took place in a tiger reserve, though they do not study tigers.) Three students from WII joined Chellam and Kumar to do their own Ph.D. work on reptiles, amphibians, and small mammals, respectively.

In the dense rain forests of southern India tidy experimental projects sometimes come crashing into the reality of work at the very edge of scientific knowledge. The work of Divya Mudappa, a WII researcher working on small mammals in the Kalakad-Mundantharai Tiger Reserve turned up some real surprises, not least because work on small mammals is simpler to write about in grant proposals than it is to actually do in the field. Compared to the larger mammals like tigers or ungulates, small mammals in India have been ignored by most hunters, wildlife enthusiasts, and biologists. Before Mudappa's project began, no one was even sure what small mammals lived in the Kalakad rain forest or where they would be found. Mudappa first did fecal collection and camera trapping throughout several rain forest fragments and at Kalakad. She found she could not distinguish among the fecal samples. [71] She could tell that there were mammals in the forest, where they were, and what they ate, but she could not match that data to any specific species. Camera trapping, which uses bait attached to a trip wire that sets off a flash camera, allowed her to identify which species are present in a given area. This was particularly useful for determining the presence of nocturnal mammals, but it did not tell her much about their habits, other than where they walk and which baits attract which species. Collectively, these methods gave her an indication of the approximate abundance of small mammals in the rain forest, at least for the species attracted to her traps. In fact, Mudappa was flying blind, largely working on her own, with no precedents to assist her. What did the animals eat? When did they eat? What were their activity cycles (day and night)? Were they rare or abundant?

The Kalakad rain forest is a difficult, though scenic, place to work. It rains there nearly every day and even in the middle of the day it is so dark that a flash is needed for any photography. Because the forest is in the mountains, it is cool at night even in the middle of summer and the rain can be cold. Working in a dense forest on large mammals can be difficult, as Christy Williams had demonstrated with his elephant project. Working in a rain forest on small mammals whose very survival depends on their ability to elude larger predators is orders of magnitude more difficult. There is a good reason why biologists have more frequently studied large mammals—it is easier to find them, watch them, or even dart them. Much is made of charismatic megafauna, but small mammals can be awfully cute, and if they could be found, they too would sell well on TV back in the United States. Mudappa's camera-trap photos are among the only live photographs that exist of some of the animals she studied. [72]

After much preliminary work Mudappa decided to investigate three small mammals that appeared to be fairly abundant in Kalakad, the brown palm civet, the brown mongoose, and the Nilgiri marten. Practically nothing is known about these species. She radio-collared animals from April 1998 to December 1999. She found, for instance, that the civets seem to have a very specific home range that is not very large.

It was not possible for Mudappa to collect the refined data originally

Figure 11. Deep inside the rain forest of Kalakad-Mundantharai Tiger Reserve is a Wildlife Institute of India field station, home to two graduate student researchers and their field assistants. The building was once part of a cardamom plantation, and much of the forest in the background hides former cardamom fields.
Cassandra Kasun Lewis, May 1998

envisioned by Chellam and Kumar or even be able to make estimates. She tried to do mark-recapture tests with rodents to estimate overall population size, but the population density was so low that she had no success. (She marked some rodents, but never saw them again.) Mudappa nevertheless continued to chip away at the edifice of mystery that surrounds the smaller creatures of the rain forest.

Mudappa's work illustrates one of the key difficulties Indian ecologists face when conducting natural experiments. Even elementary knowledge of many species is absent. G. S. Rawat, along with Johnsingh and Goyal the first three biology faculty at the WII, elaborated that until there is more baseline data on Indian wildlife, experimentation is difficult: "I strongly believe that in ecology we should try to do lots of experiments, but some of our projects are exploratory in nature. We have short-term projects. . . . Quite often we end up producing status surveys, and general documentation of vegetation, flora, and some animal and plant communities around. . . . If we had restrictions and could not work in large areas, maybe if we were limited to small, small fields, then we could do experiments."[73] Chellam and Kumar continue to receive money from U.S. grant agencies at least partly because their proposal conformed to expectations that the data would be drawn from experimental research. Chellam still hopes to be able to use the amphibian data to "predict some, or understand, patterns of extinctions based on theories of island biogeography." But he recognizes that the small mammal study is not experimental. Instead, it will produce "a whole body of new [observational] information about these small carnivores, so at both the species level and community level, a lot of baseline ecological information is going to become available." What looks good to funding agencies and what is actually needed or doable in the field are not always the same things. It seems certain that this mismatch is one of the effects of researchers in India applying to funding agencies in the U.S. for the funds for their research. Indian ecologists are placed in the position of writing proposals conforming to scientific goals or methods that are not necessarily feasible in their local research programs. And perhaps this perpetual exercise of writing projects that never get carried out exactly as planned leads some Indian wildlife biologists to assume that U.S. ecologists do not also have the problem of trying to appear to be more experimental than is possible in fact. As Hubbell would tell them, even in the United States there are many ecologists who find themselves in the same position. Hubbell, though, does not idealize a distant ecology practiced by others that *does* meet a purportedly higher standard.

Technological Fixes

One explicit goal of the WII was to adapt U.S. research technology to the Indian context.[74] The assumption behind such a goal was that technology can be, and is, adapted to local needs and situations without necessarily carrying any accompanying cultural baggage.[75] The scientific technologies sought by the founders of the WII were fairly specific in purpose, though— radio collars and transmitters track the movements of animals. While the information derived from such technologies can be adapted to different uses, it does produce a particular type of knowledge about animals, unlike any older means of observation. For some researchers at the WII the use of sophisticated technology gained them entry into a realm of scientific discussion that would have otherwise been closed to them. It also helped separate the science conducted by the WII from work done outside the WII, while it also validated their work to the editors of international journals.

For example, S. A. Hussain, who worked on otters for his Ph.D. at the WII in 1987, got most of his reading through his correspondence with others who worked on otters, because it was not available in India. His foreign helpers in the U.S. and Britain eventually supplied him with radio transmitters that could be implanted in otters (the technology is too expensive for most developing countries), and they trained him in how to use them. This implant technology, an American innovation, generated the data around which Hussain's thesis revolved and has changed the collection of data for many researchers.

Hussain's dissertation on otters was a success and he was subsequently appointed to the IUCN otter group. At the 1998 International Otter Symposium in Prague he was asked to chair a session on radio telemetry in otter studies, where the audience was very receptive. "I had underestimated my knowledge. . . . They [the British and Americans] developed me and now they are relying upon me to transfer my knowledge. It goes back and forth."[76] In addition to the Prague conference, Hussain had traveled twice to Thailand to train scientists in radio telemetry for otters.

For Hussain, expertise in this rather specific and rare technology—implantable radio telemetry—has translated into a position of respect in the global community of otter scientists. While he acknowledges that he learned this technique primarily from Americans, he now is in a position to train other First World scientists, as well as to spread the technique within Asia and the rest of the world. As he modestly noted, he has come a long way from his childhood in a rural village without electricity, with thatch roofs, and schooling in his native Oriya language.[77]

Hussain also admitted that he was able to have access to and learn to use this technology because of his placement in an institution with U.S. collaborations—implantable radio telemetry is expensive. Divya Mudappa, the small-mammal biologist working in the rain forests of Kalakad, told me that she thought the only way that U.S. collaboration had affected her project was through the use of radio telemetry. R. Sukumar, who did his Ph.D. at the IISc on elephant-human interactions, did not have access to radio collars. Admittedly, Sukumar was working in the late seventies and early eighties, at a time when no one in India was using radio collars, but even now no one at CES uses radio telemetry. Christy Williams, who had to conduct part of his research on elephants in Rajaji without radio collars, anticipates being able to accomplish much more, and more thoroughly, with them. Without the collars, he would be forced literally to spend all day looking for an elephant, sometimes with no success, and it is difficult to get natural history data without seeing your animal. Similarly, it is easy to imagine how difficult Mudappa's project would have been without radio collars.

Radio telemetry is a powerful tool for ecological research. With radio collars scientists can determine animals home ranges, migratory routes, activity cycles, breeding timing, and even how often they interact. Radio telemetry does lead to certain research subject biases, though. By their very nature, studies using telemetry are much more likely to focus on animal movement, and they are therefore much more likely to support conclusions about the necessary size of protected areas for collared animals. Single-large arguments, remember, are frequently based on the range requirements of free-moving animals. Conversely, studies used as evidence for several small reserves are almost always finely grained observational studies of the biodiversity of very specific (small) areas, which illustrate how much actual biodiversity *is* preserved in small reserves. I do not mean to caricature radio telemetry studies as just being about movement and range size, though. Williams, for instance, does do more than just track elephants, he notes their activities, group composition, and what they eat.

New technology is making possible fieldless fieldwork that would eliminate the possibility of combining radio telemetry and observational ecology. It is now possible to obtain radio collars that have built-in computer chips that periodically record animals' locations via GPS satellites. The researcher can then download all the acquired movement data remotely from a computer at home or at the office. Data from this kind of technology is limited to the study of animal movement, again primarily supporting conclusions about needed reserve size. Finally, radio telemetry, though the device can be implanted in an otter and even attached around a large frog's waist, does have size limitations and this constraint might influence a researcher's selection of study animals. There is already an animal size bias in most wildlife research; radio telemetry reinforces this. As seen in discussing SLOSS, studying large animals usually leads scientists to support large reserves, while conversely, scientists studying small animals (insects or amphibians, for example) are more likely to accept, if not support, multiple smaller nature reserves protecting a variety of microhabitats. In spite of these concerns, if I were doing work on a subject that could benefit from telemetry I would want to use the technology to generate *some* of my data. I just question whether the technology transfer is as value free as scientists and policy planners assert. I think radio telemetry clearly produces data that are easily adapted to policy recommendations for large nature preserves.[78]

Beyond any project design biases that might develop by the use of new technology, for most researchers cutting-edge technology gives their work the stamp of international approval as solid science. Y. V. Jhala, who earned his Ph.D. at Virginia Tech after winning a fellowship to a Smithsonian training course and subsequently meeting a professor from Virginia Tech there, conducts the most technologically sophisticated research at the WII. Jhala's research on Indian wolves is funded via grants from the USFWS, National Geographic, Earthwatch, National Fish and Wildlife Foundation, and the Conservation Treaty Support Fund, all U.S.-based organizations. For his latest twist on his wolf research he is dosing wolves with stable isotopes of heavy water and then recording its gradual loss to calculate energetics (the use and flow of nutrients) for the wolves in the wild. To do this, he needs heavy water, which costs five thousand dollars per liter, and he needs special recapture collars that contain a computer chip that the researcher can remotely command to fire a dart. The dart puts the wolf to sleep so that the researcher can then work on the animal without any stress to the wolf (or the researcher!). These collars cost about fifteen hundred dollars each. Jhala is getting the first batch made since they started being manufactured after a ten-year absence from the market. Jhala resists the idea that his research is any better by virtue of its technological sophistication, however: "if you want to do science, it just requires brains. You can do it with a pair of binoculars if you set your hypothesis right, and your goals are clear. . . . If I didn't have the money I could still do science."

Because of his exemplary fund raising, which he thinks is certainly helped by his U.S. education, which developed his ability to present his ideas, Jhala can pursue his interest in wolf energetics. And while he believes that he could do science with nothing but binoculars, it is also probably true that his research has been more readily published because he uses heavy water and recapture collars. Both binoculars and telemetry might be equally good scientific methods, but technologically sophisticated studies appeal to the editorial boards that control the top international journals. Trendiness matters, and technology is trendy in ecology (and in twentieth-century science as a whole).

A Theoretically Informed Wildlife Biology

By far the most common method by which the scientists and researchers of WII attempt to join international ecological conversations is through designing their (essentially baseline data collection) research in a theoretically sophisticated fashion. This means drawing upon the quantitative evolutionary-ecology tradition, first brought to India by Madhav Gadgil, which dominates most ecological publications in the West. Observational ecology can easily be tied into such a theoretical framework, even without modeling (which is rare at the WII). This is most easily illustrated with an example.

In 1994 the World Bank launched an "ecodevelopment" project for the Great Himalaya National Park (GHNP), providing $1.75 million over a five-year period, to be divided between three sectors—park management, community initiatives around the park, and research conducted by the WII on the park's flora and fauna.[79] The WII put together a team of seven researchers to work in the GHNP, on mammals, birds, plants, insects, socioeconomic factors, grazing, and remote sensing. One of those researchers, Vinod T. R.,[80] allowed me to accompany him on one of his week-long study hikes into the mountains. The World Bank ecodevelopment project included a goal of obtaining baseline biological data on the park. To that end, Vinod recorded any mammals he observed, how often he found them, and where they were. The GHNP is a relatively new park and its records are very sketchy. Vinod provided baseline data on which mammals are present and in what abundance—though he admitted that his data collection in this regard was somewhat restricted. He was not able to be comprehensive and survey the entire park. These data were all that the WII and World Bank demanded of him, and they were the data that were immediately needed for park management. The same research earned him his Ph.D. degree.

Vinod chose to go beyond those goals, however. He conducted in-depth research on three ungulate species found in the park—the goral, the Himalayan tahr, and the musk deer. Further, rather than just presenting life history data on these species, Vinod analyzed their distribution to test theories of ecological separation or niche partitioning. Specifically, do the three species inhabit different ecological niches, as separation theory requires, so that they avoid competing for similar resources? How are the animals distributed in space, and what are possible explanations for that distribution? Vinod clearly was motivated to do work that is ecologically significant and that would contribute to the way scientists throughout the world think about resource competition and niches among copresent animal species. Thus, Vinod was taking what might be perceived as mere stamp collecting, a field taxonomic study, and was turning it into a theoretically informed piece of ecological research.

Vinod's expansion of his work to test a theoretical model represents something important. The biologists of WII were originally appointed for

Figure 12. This picture of some of the tall peaks of Great Himalaya National Park was taken at ten thousand feet, moments before a snowstorm drove us into a small shelter for the day. The clearing that offered such a spectacular view exists, in large part, because it was often grazed during the summer months. *Photograph by author, April 1998*

utilitarian purposes. They were intended to conduct baseline research about endangered animals and habitats and to pass on the latest in management theories to forestry trainees. Instead, the WII biologists have turned the tables and are using the institute itself as a means to gain funding for their research. Vinod satisfied the requirements of the World Bank, but his true energy and attention were directed elsewhere, much as Schaller and Spillett satisfied the requirements of the Johns Hopkins Public Health program that funded their work. Vinod wrote two reports when his fieldwork was done, one for the World Bank and one for his Ph.D. committee. Vinod identifies himself as an ecologist, and his projected audience for his work was other ecologists throughout the world.

This shift in focus leads to possible conflicts. The government of India often commissions WII researchers to conduct brief surveys in areas where there is some environmental conflict. The WII reports are then the official reports of the Ministry of Environment and Forests. It is not difficult to imagine a scenario in which a researcher could write a report that leads to recommendations contrary to the ministry's goals. If the researcher identifies herself as primarily a research ecologist, versus a management scientist, she might be less willing to change her (internationally publishable, but politically unpopular) report under pressure from the ministry.

Such a mess is not what the Ministry of Environment and Forests envisioned when it set up an autonomous institution to gather baseline ecological information. Since I was in India, many of the younger biology faculty have left the WII, unhappy with its restrictions.

The WII, along with the BNHS and CES, completes a triad of institutional structures in India that promote ecological studies that are integrated into U.S. systems of funding, personnel, methods, and theoretical bases. The WII also represents the resumption of significant U.S. involvement in Indian ecological studies after a decade of exclusion. While the CES was largely begun without direct U.S. technical or financial support, the WII resumed the older pattern established with the BNHS in which U.S. governmental and nongovernmental organizations collaborated with Indian scientists and funded their studies. Unlike the BNHS collaborations, however, collaborations with WII faculty are organized in such a way that the U.S. scientists serve only in advisory capacities and do not carry out the fieldwork themselves. Further, individual U.S. universities do not collaborate with the WII, except insofar as they agree to train WII faculty on USFWS- or FAO-supported study trips.

The WII is also of interest because of the way in which the biology faculty and graduate students have positioned themselves in international discussions of wildlife biology, a move not foreseen by the institute's founders. U.S. scientists and students who are working in conservation biology mirror many of the concerns expressed to me by researchers and faculty in biology at the WII. The U.S.-based Society for Conservation Biology's goal is to encourage the application of ecological knowledge to conservation—a goal very similar to the hopes expressed by most WII researchers. U.S. scientists who work in conservation biology are often confronted with the same problems of experimentation, technology, and funding, and of turning observational studies into theoretically sophisticated work.

Because of India's position as a mega-diversity nation with a largely unstudied flora and fauna, and given the context of the current international environmental movement, the WII is often well-positioned in grant funding competitions. This does not mean that WII biologists do not have to work hard to scrape together money for projects, or that they are not rejected by more granting agencies than fund them. But compared to other institutions in India, or throughout the world, the WII is not doing badly. The funding for these studies results from two impulses in conservation biology, both originating in the U.S. On one hand, there is funding, like that for Vinod's project in GHNP, for baseline flora or fauna surveys that arises from real anxieties about inventorying diversity in an age of extinction. On the other hand, there is funding, like that for Chellam, for work that conforms to current ecological trends (such as manipulative experimentation, or the use of sophisticated methods) and is traceable to a respect for active scientific interventions. In both cases, the researchers of WII have been and are able to adapt their proposals to secure funding for their projects. In so doing, the biologists are pushing the international discussion of India's protected areas and protected species further away from the IFS and issues of governmental management.

In the early seventies, the IFS and the Government of India officials concerned with conservation policies reasserted national control over the practice of ecological science in India. In the most charitable reading of their actions, they feared the influence of U.S. agendas and the return to the days of colonial science, in which knowledge about the nature of India was created and appropriated by outsiders. They also saw the high international profile of conservation initiatives and wanted the Indian government, and Indian actors, to get the credit for Indian conservation policies. Men like Ranjitsinh and Sankhala wanted to save the tiger—they were passionate about conservation in their homeland. It galled them to think of people from the United States, or Britain, potentially remembering Project Tiger as an example of "the West" intervening to save India from the Indians. And though I have been critical of the science of Project Tiger in this chapter, it has been in many ways a conservation success.

Less charitably, though, the government of India and the IFS could be seen in this period as attempting to reassert government control of a significant portion of the Indian landmass, away from the potential control of Indian ecologists—people like Ali or the Indians who ran the WWF-India chapter, people who might have a different agenda than the government. Missing from my chronology of Indo-U.S. relations thus far is one other event of that decade, the Indian Emergency. In 1976–77, Prime Minister Gandhi imposed dictatorial rule on India for a short time. Among the governmental excesses of this blessedly brief period, including forced sterilizations and the bulldozing of slums, we can see the same confidence in Indian governmental expertise that characterized the decision of Ranjitsinh and Sankhala (with the prime minister's blessing) to go it alone on Project Tiger just a few years earlier. And it was only when Indira Gandhi was voted out of office for two years that the thaw occurred that resulted in the resumption of U.S. ecologists working in India, and the resumption of PL 480 funds being available for research. When she returned to office in 1980, she did support the idea of the 1982 Indo–U.S. Conference, but the crucial resumption of collaboration had occurred while she was not in power.

The WII was the most tangible result of the resumption of that openness to ecological collaboration, though it was still placed firmly under governmental control. As consistent Indo–U.S. collaboration returned, as well as international collaboration represented by funding from the FAO and UNDP, another group of Indian ecologists emerged. Their first loyalty was not to the government or to the IFS, but often to conservation and science—this is especially so among the graduate students. These new biologists, the WII biologists and graduate students, are now positioned in a Ministry of the Environment and Forests institution, though (not in a NGO like the BNHS or an independent institution like the CES), and their science is officially presented as a source of governmental information to be used for management decisions. Although this is advantageous for biologists hoping to get access to India's protected areas, it leaves open the possibility of the government attempting to assert control over their research.

CHAPTER EIGHT

All Nature Great and Small

Designing Indian Nature Preserves

> I am sure that the authorities in India who have been working hard for the establishment of national parks . . . would be quite unhappy with the conclusion . . . that the establishment of the national parks should be undertaken with caution.
>
> —*Lee Talbot, 1969*

NATIONAL PARKS IN INDIA GIVE RISE to many debates. There are debates about people living in national parks—as with Williams's elephant studies in Rajaji National Park. There are debates about people collecting nontimber forest products, like mushrooms, honey, bamboo, or grass for fodder. There are debates about how to draw park boundaries—how big they should be, what they should or should not include. There are debates about how national parks should be managed and by whom.

Two broadly defined debates in particular demonstrate the impact of conservation science and scientists on the management of India's protected areas and, by association, on the hundreds of thousands of people who live in and around India's national parks.[1] These debates revolve around how to define "natural" in national parks (e.g., do cattle or people belong?), and how big a national park needs to be in order to be effective (the SLOSS debate, reprised). By this point, you are aware of the links between U.S. and Indian ecologists. Here we can see the impact of these scientists and their advocacy on India itself.

Letting Nature Take Its Course: People, Cattle, Buffalo, and Birds

Between Sálim Ali's first visit to Bharatpur in 1935 and his death in 1987, he oversaw the transformation of these beloved wetlands from a private hunting ground to a national park. Just months before it was declared a national park in 1981, the BNHS began a ten-year research project on the ecology of Bharatpur that stands as one of the most important Indian ecological studies of the last thirty years. The scientists hoped to learn how to best manage the wetlands to maintain diversity; an unspoken hope was to show how dramatically the wetlands would improve as a functioning ecosystem following the implementation of the policies resulting from Bharatpur's transition to a national park. After the new designation, surrounding villagers were prohibited from collecting fodder for livestock and grazing their cattle and buffalo within the park boundaries. The wetland became more "natural." The results of the BNHS study were a shock to ecologists throughout India and have been a severe challenge to the understood methods of national park management in India.

Bharatpur's History

There are not many wetlands like those at Bharatpur left in northern India. Throughout the world, wetlands have been among the least protected of natural ecosystems.[2] There are several good reasons for this. Local populations often perceive wetlands as places of danger. They breed malarial mosquitoes, poisonous snakes, and larger dangerous animals (in India, this includes the aptly named "mugger" crocodiles, tigers, rhinoceroses, leopards, and elephants).[3] One of India's most famous wetlands, the mangrove swamps of the Sunderbuns, averages between forty and sixty human fatalities a year from tiger attacks. Beyond dangerous wildlife, wetlands are difficult terrain through which to travel and are prone to such catastrophic inconveniences as flooding. In contrast to their danger when wild, once tamed and drained, wetlands are among the most productive agricultural lands available. The hundreds of millions of people who live in northern India are fed by the wheat, rice, and sugarcane grown on the former wetlands surrounding the Ganges and its many tributaries. Western-style "development" has been, and continues to be, experienced in many parts of the world as attempts to control, regulate, and distribute water through damming rivers and draining wetlands.[4]

The scarcity of wetlands in Northern India has resulted in a much higher concentration of birds at Bharatpur than would be considered natural in a less modified, fragmented landscape. But then very little about the national park there is "natural." There are no top predators in the park—a tiger wandered into the park last year but it is uncertain how long it will stay, and the last leopard was killed over twenty-five years ago. There is a very limited assortment of nonavian fauna in the park. It does not have a natural ecosystem. More directly, the wetlands now protected in Keoladeo Ghana National Park are not the result of erosion, plate tectonics, or drainage patterns, but rather are the product of dams and dykes constructed on the orders of the maharaja of Bharatpur in the 1890s.

The maharaja was not interested in bird conservation for its own sake. On the contrary, he ordered the construction of the wetlands when he had just returned from a trip to Great Britain, where he had greatly enjoyed the waterfowl shooting excursions, and he was determined to have a shooting reserve for himself.[5] He selected a small preexisting marsh and expanded and deepened it to make it more attractive for wintering birds and year-round bird residents, using many canals and dykes to regulate the water level. He then crisscrossed the area with walking paths so that hunters could have access throughout the wetland. You can still walk on those paths, some still paved with bricks placed there one hundred years earlier, but now visitors to the national park wield cameras, not guns.

The maharaja was successful beyond his wildest dreams. Bharatpur became famed for its bird hunting and the maharaja entertained hundreds of visiting British and Indian dignitaries. Stone plaques in the middle of the national park that remain from earlier, bloodier, days record visitors and their bags—the listings include Lord Curzon (then viceroy of India, the first of many to visit Bharatpur), almost all the local maharajas of northern India, the Prince of Wales, the crown prince of Germany, and many others. Their daily kills were stupendous, sometimes exceeding four thousand birds.

In spite of these hunts, there seemed to be a never ending supply of birds for Bharatpur. As frequently as ducks were shot from the sky above Bharatpur, wetlands in the rest of northern India were being drained and converted to agricultural land, driving birds wintering in those places to the remaining wetlands. Within northern India the conversion of wetlands to agricultural land is not a recent process, but it accelerated in the twentieth century and especially after India gained its independence and embarked on a national push for agricultural self-sufficiency.[6] In a country of dwindling wetlands, Bharatpur became increasingly unusual as a protected and stable habitat for waterfowl.

When India achieved independence, in 1947, local maharajas signed agreements with the central government transferring power from their states to the center. When the maharaja of Bharatpur negotiated his deed of transfer of power, he made sure that he retained ownership and the "exclusive right of shooting in the Ghana [wetland] for himself and his friends."[7] Once the initial heady rush of independence had passed, local people who had been the maharaja's former subjects began to protest his continued privilege in reserving land for his personal pleasure. They argued that this was a clear case of a royal shooting reserve serving no purpose other than the ego gratification of a former ruler and his friends and that it denied poor landless farmers in surrounding villages land and water. There were no actual rights among the surrounding villagers to the resources of the wetlands, although local cattle and buffalo were allowed into the reserve on a day-to-day basis for grazing. Bharatpur, local politicians argued in 1950, should be converted to agricultural land for landless laborers, and its canals should stop diverting water from other agricultural areas.[8]

Ali, then head of the BNHS did not see this debate between the maharaja and his former subjects as centered on questions of the abuse of privilege. He saw Bharatpur as a threatened wetland, one of the last remaining wintering grounds for water birds in northern India. His sympathy was not with the poor but with the birds. He saw the antimaharaja forces as motivated by personal politics and as ultimately evil, that they were "prepared to go to great lengths to punish him for alleged misdeeds and injustices in the past. They had whipped up an agitation alleging that he was holding on to the Ghana solely for his selfish pleasure and to entertain influential friends at court in Delhi, and thereby wickedly depriving the poor land-starved ryots [peasants] of good cultivable land and the water they so badly needed for their crops. With the backing of crooked politicians complete physical devastation of the Ghana was plotted."[9]

Ali wasted no time in protecting the wetlands—he went straight to his personal friend Jawaharlal Nehru, the prime minister of India. Nehru had his minister of agriculture and irrigation check into the charges of land and water deprivation. The government found that Bharatpur supplied many needs for surrounding poor villagers, including fodder, firewood, berries, and thatch grasses, and would be of more use to poor peasants continuing as it was than if it were converted to agricultural land. Bharatpur was thus "saved from certain annihilation."[10]

However, in 1956 the wetlands were officially turned over to the Rajasthan Forest Department for management and maintenance.[11] Bharatpur was saved from agricultural conversion not on the basis of the Maharaja's rights to the land as personal property but on the basis of its biological resources and use value for surrounding villagers. To prevent future conflicts between the local populations and the maharaja, Bharatpur was turned over to the state. The maharaja was able to retain hunting rights for another few years, but since 1968 the echoes of gunshots have not sounded across the marshes.

That is not strictly true—no more shots have been aimed at birds. Bharatpur was declared a national park in August 1981, and so it officially became subject to the 1972 Wildlife (Protection) Act, which required that national parks be areas free of human activities and of livestock. The law made little difference initially. Surrounding villagers had depended on the vegetation of Bharatpur for decades as fodder and thatch—that was how, after all, the central government had justified not converting the wetlands into agricultural land in the early 1950s—and villagers continued to use Bharatpur for grazing after 1981, in spite of the letter of the law and a large stone wall built around its perimeter. They would simply enter through the gates, unstopped by the local forest guards. In October 1982 the Indian Board for Wildlife held a meeting in New Delhi, chaired by Prime Minister Indira Gandhi, that addressed the problem of grazing and human pressures in Bharatpur. The board decided to enforce the rules and completely close the park to grazing, and further, that this decision was to be implemented right away.[12] Gunfire returned to Bharatpur following that fateful decision and nine villagers who were trying to enter the park during resulting protests were killed by police.

Cattle and Conservationists

Cattle and other domesticated livestock inside protected areas have been the bête noire of many ecologists, environmentalists, and wildlife preservation managers who have worked in India. This has been true in other regions of the world as well— throughout his life John Muir famously referred to sheep in the Western United States as "hoofed locusts." Calls to eliminate domestic animals from wildlife sanctuaries are repeated time and again in policy documents, scientific papers, conservationist writings, and popular articles. Kailash Sankhala of Project Tiger sees cattle grazing as the biggest problem threatening Indian wildlife and the national parks. The Indian Board for Wildlife, in its recommendations on Wild Life Sanctuaries in 1965, recommended the prohibition of grazing of domestic animals in sanctuaries.[13]

George Schaller was more nuanced in his discussion of cattle in *The Deer and the Tiger*. His first mention of cattle is harsh: "A great scourge of India's land is the vast numbers of domestic animals which are undernourished, diseased, and unproductive, yet are permitted to exist for religious reasons."[14] At the end of the book, though, he is more moderate. Based on his careful study of the tiger's feeding habits, he concludes that livestock grazing within Kanha National Park helped sustain an abnormally high density of tigers. Specifically, he claims that tigers killed enough livestock (cattle and buffalo) in the park each year to completely fill the dietary needs of ten tigers (250 head of livestock). "From the standpoint of tiger conservation," he writes, "it makes no difference whether the prey consists of cattle or wildlife" (328).

He was not just interested in tigers, though, and "from the standpoint of habitat conservation and the maintenance of the park as a sanctuary devoted to the perpetuation of wildlife, the livestock should ideally be eliminated and the wild hoofed animals be permitted to increase" (328–29). Ultimately, therefore, park managers should remove the livestock from within the park. Only then, "will the park be able to fulfill its unique potential as a living museum and natural laboratory. Above all, Kanha National Park is part of India's cultural heritage, a heritage in many ways more important than the Taj Mahal and the temples of Khajuraho, because, unlike these structures formed by the hands of men, once destroyed it can never be replaced" (331). Schaller was seeking to add the 319 square kilometers of Kanha National Park to the rolls of living museums—places dedicated to the preservation of earlier times, earlier ways of life, and in the case of his view of a national park, prehuman ecosystems—while thousands of people living in (at that time) and around the park continued to experience their life in the park as very much the present, and its nature as in no way prehuman. This language mirrors that of many Indian environmentalists who speak of the need to preserve "India's precious natural heritage," or of Project Tiger promotions that attempt to equate India's nature with Indian uniqueness. This language reinforces a particular vision of the Indian nation and it does so at the expense of local populations like those who herd the cattle of Kanha.

By far the most thorough critique of livestock in Indian national parks was written by Juan Spillett, in his summary of the state of Indian wildlife preservation prepared for the BNHS. In an article entitled "General Wild Life Conservation Problems in India," Spillett wrote that India is basically confronted with two major problems: "I firmly believe that if these two were brought under control, the numerous other problems which are presently receiving so much attention and publicity, such as the scarcity of food, lack of foreign exchange, poor living standards, and so forth, would eventually resolve themselves. . . . These two problems are: (1) too many people, and (2) too much domestic livestock."[15] This statement directly links poor Indians with cattle as the scourge of India. Spillett follows this astounding claim by comparing livestock grazing to a cancer that quietly destroys an ecosystem, no one noticing until it is too late. India should know better, he claims, for "primarily due to over-grazing by domestic livestock, India already has the notoriety of having created the largest man-made desert in the world." He adds that "more nations have fallen because of land abuse . . . than by all other factors combined."[16]

Such arguments are questionable but not unusual. Vasant Saberwal has traced debates blaming desertification and erosion on livestock grazing in India. He found that scientists and bureaucrats consistently make the (flawed) assumption that overgrazing causes deserts. In fact, Saberwal asserts that there is no direct correlation between grazing and desertification in India and that this sort of monocausal analysis is misleading.[17] It is a powerful environmental myth, though. We all have seen the photographs of a bare-ribbed cow standing in the midst of a sand dune, plucking the leaves off of the only plant in the frame. Such presentations are often more effective in shaping the popular imagination than arguments that global climate change and weather patterns cause deserts, not overgrazing. To this point, grazing has destroyed no park in India, even though for nearly half a century conservationists have feared that it will happen.

Misidentifying the causes of deserts is a mild sin. Far more severe is making policy recommendations based on those beliefs that result, literally, in human deaths. Spillett states,

> I am almost invariably told by officials that [they realize] the problem . . . , but that it is impossible to control grazing by domestic animals in a democracy such as India's. This is faulty reasoning. No government, particularly a democratic one, should permit its people to destroy the nation's most priceless possession—its land. Many feel that in a democracy public property belongs to everyone. But this does not mean that the people are free to destroy the public domain. For example, a public building belongs to everyone just as much as does a reserved forest or a wild life sanctuary. However, no one is allowed to destroy such buildings or to remove materials from them for private use.[18]

Spillett provides here the justification for treating livestock grazing in protected forests as a crime equivalent to terrorist bombings of buildings or
theft. He misunderstands the problem of democracy posed by the officials
with whom he spoke; he assumes that the officials are claiming that you
can't force people to do something like stop animals from grazing. He argues that, just as you would shoot a terrorist before he blew up a building, a democratic state would also justify stopping a herder from taking his
cattle into a protected area. The problem of democracy in India's national
parks, though, is a different one. The vast majority of people living around
Indian national parks—and perhaps in the nation as a whole—disagree
with park policies prohibiting grazing and human use. They do not see it
as a crime. The people living around parks are not elites, though, and their
voices are often absent in environmental debates. And arguments such as
Spillett's allowed the Indian Board for Wildlife and Indira Gandhi to make
the decision to exclude grazing, by force, at Bharatpur. Livestock grazing,
after all, was a crime with consequences to the Indian people, they might
have stated.

The Spillett, we must remember, was not working on his own when he
wrote these words. He was a graduate student writing on the behalf of the
BNHS, lending his American citizenship and training to their crusade. As
the editors of the *JBNHS* had written in their introduction to the volume
in which his articles were published, "It is not irrelevant to refer to
Spillett's conclusion, that it is not only poaching that is the great threat to
the continuance of India's wild life and wild places, but the disastrous
effect of over-grazing by domestic livestock all over the country. Understanding a problem clearly is the first requirement toward its solution, and
we are grateful to the author of this report for the facts which he has un-
covered."[19] Spillett was not a lone voice in the wilderness, neither was he
a ventriloquist's dummy, mouthing words he did not believe. He was given
the platform of the *JBNHS* because the BNHS staff agreed with him and
thought he could further their aims.

The BNHS consistently relied on nondemocratic politics to effect
environmentalist goals, as when Ali approached Prime Minister Nehru
about the maharaja's threatened shooting grounds. In a 1983 interview Ali
said, "Years ago when I could have used my friendship with Jawaharlal
Nehru to obtain 'favors,' . . . I hesitated, [thinking] that he probably had
enough problems to contend with as it is. How wrong I was. . . . Anyhow,
I certainly shan't repeat the mistake where his daughter, Indira Gandhi, is
concerned."[20] Perhaps the intervening years had obscured Ali's memory of

his and Ripley's several appeals to Nehru—about Bharatpur, about the Gir lions, about Ripley's *New Yorker* article, among others. Or, perhaps he felt that his previous history with Nehru truly was hardly anything, and he truly intended to be even more direct with Gandhi. In any case, Spillett's justification of action based on an elite conception of the public good was in keeping with the preferred mode of environmental advocacy by the BNHS and most other international environmental organizations working in the developing world. The World Wildlife Fund very directly claimed that one of its goals was to approach national leaders directly and affect top-down environmental action, as when Prime Minister Gandhi was persuaded by high-level WWF members to support Project Tiger.[21]

Within a year of Spillett's article for the BNHS, there was a flurry of cattle-based activity in the BNHS and the Smithsonian. In 1967 the government of India established a parliamentary committee "to investigate the implications of a total ban on the slaughter of the cow and its progeny."[22] Conservationists in India and throughout the world took note. Zafar Futehally wrote an editorial page article for the *Times of India* in which he decried the effect of cattle on the Indian landscape, and in true BNHS fashion, he called on the international experts: "In our own sanctuaries in India, as reports by George Schaller and Juan Spillet [*sic*] show, much harm has been done to wild life and to the habitat in such unique areas as Kanha, Kazaringa, Sariska, and Bharatpur."[23] Futehally was serving as the Indian chair of the International Biological Programme Terrestrial Conservation subgroup, and he moved to get the IBP to fund a study in India on cattle and wild herbivores. He approached the Smithsonian to check on the availability of PL 480 funds. The Smithsonian was amenable but the parliamentary committee needed to submit a report by the end of the year. Although the committee was willing to listen to any scientific advice, there was no time for a complete study.[24]

This called for direct political action—so Ripley wrote a letter to Prime Minister Gandhi, asking for her to intervene so that a rupee-funded study could take place (note that this whole sequence of events was set in motion by Futehally and the BNHS, not by outside conservationists). Ripley received no response, until finally the U.S. ambassador to India sent a deputy to check with one of Gandhi's assistants. The ambassador reported to Ripley, "Haskar readily confirmed that it had been received, touched briefly on the sensitivity of the subject, and as much as said he thought it better to leave the complexities of the cow problem to the Government of India." The ambassador told Ripley not to expect a reply from Gandhi and, further, suggested that in the future Ripley go through the U.S. State Department if he wanted to contact the head of the Indian state. In a gentle rebuke, the ambassador closed by suggesting that Ripley learn what the Indian government thought were their major problems (as opposed to cattle): "the modernization of agriculture, and the control of population."[25] The files in the Smithsonian Archives usually include attached articles, clippings, and memoranda that come with the letters found in those files, no matter how mundane, but those two memoranda are missing. Might Secretary Ripley have thrown them away?

Only two years later, a team of graduate students completed a preliminary report of an ecological study of the role of cattle—domestic and feral—in the ecosystem of the Gir Forest (last home of the Asiatic lion) conducted with PL 480 funds. It was exactly the sort of specific study that Futehally and Ripley had been pushing for in 1967. A paper by one of the graduate students, Paul Joslin, included the expected information about overgrazing and soil compaction by the cattle, all to the detriment of the forest. But then, the paper veered in a direction unforeseen by the BNHS or Smithsonian. First, it indicated that the cattle ate different plants than did the wild ungulates in the forest. Joslin then suggested that even if all the cattle and buffalo were removed, it would not lead to a corresponding increase in deer and antelope. He also proposed that the wild ancestors of the buffalo and cattle perhaps once lived in this forest and played a similar role in its ecosystem. If the cattle were gone, the wild ungulates still would not eat the newly available grass. The paper thus concluded that "removal of livestock will sharply reduce the capacity of the Gir to support lions and other large carnivores," since the lions depended on the cattle as food. "[C]urrent plans to remove all grazers and their livestock and constitute the Gir en [*sic*] toto as a sanctum sanctorum for wildlife, would mean that at most 25% (the proportion of wildlife in the diet) of the current lion population will remain."[26]

This did not go over well. Lee Talbot, a Smithsonian official overseeing international ecology projects, quickly set Joslin straight:

> I would quite agree that the sudden removal of all domestic stock from the forest would be likely to be detrimental from several standpoints, among them, the increased predation pressure that this might put on the remaining wild ungulates. However, as a long term objective, it seems most desirable to develop a sanctuary in which the wild ungulates and predators coexist as they did prior to the introduction of the present

human uses and livestock. . . . Another dimension of the problem is the India-wide problem of overgrazing by domestic livestock. The Gir does provide an opportunity for meaningful scientific research on this point and recommendations from the Gir which appear to condone overgrazing would be certainly unfortunate. Incidentally, several who have viewed your remarks on the vegetative response, its removal of livestock and its effect on the lions, have remarked that they consider you are on pretty shaky ground. I am sure that the authorities in India who have been working hard for the establishment of national parks . . . would be quite unhappy with the conclusion that can be drawn . . . that the establishment of national parks should be undertaken with caution.[27]

This is a remarkable letter. When the science Joslin had presented did not confirm the conventional wisdom, it was rejected. In conclusion, Talbot told Joslin, "I am sure it was not your intent to indicate these things . . . however, we do feel it is quite important that you make sure they are not misinterpreted in your presentation at the [IUCN] meetings." Of course it was Joslin's intention to call into question the accepted view of the need to remove the livestock! Talbot was offering Joslin a way out of this snag— after all, Joslin still needed access to Smithsonian funds to do his study, and he could not anger his funding agency.

Talbot sent a copy of the letter to David Jenkins, of the British Nature Conservancy. Jenkins reassured Joslin that Talbot's point was that "young scientists should devote themselves to their research and not become politicians. With this I wholeheartedly agree." Jenkins continued, "I am not really worried about this particular instance since I think that the final rewrite of your paper [completed between the two letters] overcame most of Lee's detailed objections." The final version, as presented at the IUCN meeting, conformed to the desires of what Jenkins called "the ecological politicians."[28] This, of course, is not how science is supposed to work.

Why Did Conservationists Fear Cattle?

There are a number of factors that contributed to the bias against livestock among many conservationists. The most obvious is the impact of cattle in areas where they are not native and have had a damaging impact (as in some places in the American tropics). Cattle are often devastating in rain forest biomes. But why dislike cattle in grasslands? Cattle also have familiarity working against them—when visitors go to national parks they hope to see unusual or at least different animals. Unusual is often glossed as "wild."[29]

The divide between a wild animal (and thus to be admired and prized if seen in a national park) and a domesticated one (and thus to be prohibited from destroying the national parks in which they graze) breaks down even further with the water buffalo. When people speak of livestock in India, buffalo are included with cattle. Buffalo are used for milk, for pulling carts (some improbably loaded with fantastic amounts of weight, including concrete and rebar in Delhi), and for working fields. Buffalo are not sacred and are thus also eaten in some parts of India (but only after their working days are largely finished). The wild water buffalo is also listed as one of India's most endangered species, found in only a handful of parks in all of India. The wild water buffalo is exactly like a domesticated water buffalo except that it is larger, with bigger horns, and (by reputation) with a nastier temper. Scientists at the WII were considering (in 1998) DNA typing so that they can look for distinguishing genetic markers between domesticated and wild water buffalo. The fear among many environmentalists is that the wild water buffalo is increasingly isolated and surrounded by its domesticated brethren and is being cross-bred to extinction. Local villagers who live near forests with wild water buffalo will sometimes let their female (domestic) buffalo breed with wild males, in hopes of improving the strength and size of the offspring. And on occasion, a domesticated male might impregnate a wild female, diluting the genetic purity of the hybrid offspring. Of such things—being bred to death—are extinctions made, in this bizarre example.

The most compelling argument against livestock in national parks and other protected areas, however, is not desertification, lack of wildness, or its sheer familiarity and lack of excitement to environmentalists. Domesticated livestock are not going to go extinct; many wild animals run exactly that risk because they cannot be domesticated and only exist in nonagricultural, nonsettled ("wild") landscapes. Given that there are a limited (and diminishing) number of square kilometers of such wild land left in the world, and that this land can maintain a finite number of animals (domestic or wild), the decision then must be made whether to dedicate that land solely to wild creatures or to allow domesticated livestock to also utilize the resources of that land as well. Almost unanimously, ecologists and environmentalists argue that wild land must be set aside for wild creatures. As George Schaller argues, tigers can't tell the difference between wild meat and cattle meat. But the presence of livestock in Kanha National Park lessens the total number of other herbivores that can survive there. The park has a finite carrying capacity, in ecological terminology. The more water buffalo or cattle that there are in the forest, the less endangered swamp deer that the forest can support.

This idea of carrying capacity combines with the lack of enchantment elicited by the familiar and mundane beasts of the world to result in a thoroughgoing prejudice against livestock grazing in national parks. As the Gir study reported, though, removing the cattle might not help the wild ungulates in the forest by opening up more grazing space. The wild animals eat different plants. So in at least that one ecosystem, the carrying capacity argument appears to be moot. The Gir ecosystem had evolved to the point that livestock use was part of its "natural" workings. Unfortunately, this information reached only a limited audience.

The Bharatpur Ecology Study

In the late seventies the BNHS (who had resumed their bird-banding project) noticed a slight decline in bird populations in Bharatpur and they could not definitively explain it. One BNHS staff member, J. C. Daniel, one of Sálim Ali's recruits from the early days of India's independence, suggested that the society conduct a ten-year study of all aspects of the Bharatpur ecology. The study was to be almost exclusively observational, recording with as great detail as possible animal populations and resource use, including the vegetational composition of the different regions of the park and fish population dynamics, and how all these elements seemed to interact. The idea met with widespread approval and was begun in 1980 with PL 480 funding (as with all these late seventies megaprojects, the USFWS was the collaborating institution from the United States).

The scientists had a decent idea why the bird populations were declining. The problem was the cattle. David Challinor, assistant secretary for science at the Smithsonian, visited Bharatpur in 1980. He was shocked, and after Ripley read his report, Ripley wrote to Ali, "there seems to be some frightful bounder who is currently running the preserve who has never completed the surrounding perimeter wall, and consequently allows quantities of cattle, buffalo, etc. to run all over the preserve mixing among the few Nilgi [*sic*] and Black Buck that are there. The cumulative effect of invasion by cattle has been to break down the bunds trampling them beyond recognition in some cases and allowing the water to escape totally."[30] The government of India had planned to erect a stone wall to keep out cattle since 1976, when Prime Minister Gandhi visited the refuge and suggested that it was needed.[31] The BNHS and other conservationists in India as well as the United States decided to advocate the declaration of Bharatpur as a national park, not just a bird sanctuary. National parks, according to the 1972 Wildlife (Protection) Act, were required to be free of cattle and people. Bharatpur was declared a national park in 1981 and, as indicated above, in 1982 the government of India (at the insistence of Gandhi herself) enforced the ban, and nine people were killed trying to enter the park in the resulting riots.[32] The Bharatpur ecology study by the BNHS, begun in 1980, was thus positioned to compare the park's ecology before and after its change in status, and before and after the grazing ban.

By 1986 the Indian environmental community had begun to whisper that surprising results were coming out of Bharatpur. A mid-study report had indicated that bird diversity in Bharatpur was dropping since the ban on grazing and fodder collection had taken effect.[33] When the final report on the BNHS project came out in 1991, it was official. A few weed species were taking over the Bharatpur wetlands, reducing the fish populations, which was also reducing bird populations and nesting. The canals were clogged, the marshes were filling up with weeds, and the grass in the grasslands proved to be dangerously productive—there had been a series of out-of-control wildfires that had not been conducive to dry-land bird nesting, and the birds had nowhere else to go. The vegetation of Bharatpur was more fecund than any scientist had anticipated, and in the absence of intense use, a few "weedy" plants (both fast-growing native species and introduced species like the grass *Paspalum* and water hyacinth) were destroying the park as an open wetland habitat so well-suited for birds. The wild ungulates in the park (nilgai, chital, blackbuck, and sambar) were not eating these weeds. In one of the truest surprise endings ever found in a scientific document, the BNHS, the strongest advocate for the notification of the wetlands as a national park and the ban on grazing in Bharatpur, concluded that "the only ecologically viable alternative [to the weed takeover] is to set the primary consumers (buffalo) back into the system."[34] To fail to do so, the report argued, would result in a continuing decline in Bharatpur's bird population, both in numbers of species found and in total numbers of birds.

The final report explained that hands-off management of wild areas would not work in Bharatpur: "In the case of most wetlands in this country, man's interaction with them has become almost inseparable and hence, man has to actively manage them."[35] The only problem was that hands-off, or "natural," management had been the international environmental vogue when India's Wildlife (Protection) Act had been written and passed in 1972 and was thus written into the law. Natural management—the strict elimination of any human "interference" with the "natural" workings of a nature reserve—means letting fires burn uncontrolled and letting animals starve to death if their population becomes too large. Advocates of natural management believed that eventually such a management style would allow the nature reserve to reestablish historically normal populations and vegetational communities.

By the late sixties many American environmentalists were convinced this was the best management style for a national park—it would allow nature to act naturally. In the United States, the 1916 act establishing the National Park Service had been passed within the context of dramatic overhunting of game species such as the American bison and its overriding concern was "saving resources from the rapaciousness of private commercial interests."[36] The act was thus open-ended as to precise policy but adamantly oriented toward preservation, not management. In 1969 the managers of Yellowstone National Park, the world's first national park and flagship of the U.S. park system, began a new management style, one of complete noninterference—natural management.[37] This policy was justified both by the original 1916 act and by scientific ideas such as "ecological homeostasis." [38] In truth this was a warmed-over version of Frederic Clements's old (and discredited) climax theory of vegetational succession, applied to an entire ecosystem—if we remove any human disturbance, the balance of nature will assert itself and the ecosystem will reach a homeostatic climax state. This policy quickly became an unofficial norm throughout the U.S. park system. The U.S. system has served as a model for the establishment of national parks throughout the world, and American parks such as Yellowstone were the model for the IUCN's international standards for national parks and other forms of nature reserves, codified in 1960.[39]

When India's Wildlife (Protection) Act was passed, just a few months after the 1972 Second World Conference on National Parks (a conference attended by the author of the act, M. Ranjitsinh), it wrote into Indian law the current U.S. national park model exemplified by Yellowstone since 1969 and explicitly advocated at the 1972 conference, of natural, hands-off, management. To see national parks in India as solely based on U.S. models is of course too simple. Ranjitsinh was from a royal family in Rajasthan, and many of the Indian maharajas had been maintaining private hunting reserves for decades, which gave them experience in managing wildlife. Ranjitsinh points to these roots when he talks about the law he did so much to write and see implemented. Similarly, British forest and wildlife reserves had been established in the late 1800s.[40] I would suggest that these earlier Indian models made it easier to accept the validity of the U.S. model of national parks. In any case, within months of a conference in which people from around the world celebrated the hundredth anniversary of the U.S. system and speaker after speaker spoke of spreading it through the world, India implemented a national park system that contained the key provisions of no people, no cattle, and with natural management.

Since 1972 natural management has proven questionable at best in U.S. parks, leading to a firestorm of criticism. In Yellowstone, the elk population, without traditional predators such as wolves, is out of control and the elk are decimating the aspen and willow stands in the park.[41] Though wolves have been reintroduced to the park, there are not yet enough of them to control the elk population. In addition to the elk, in the summer of 1988 forest fires burned 45 percent of Yellowstone, feeding on nearly a century of dry timber that had never been burned due to an earlier fire-suppression policy.[42]

The BNHS study of Bharatpur indicated that the natural management policy was similarly problematic for that park. As the BNHS found, "natural" management at Bharatpur resulted in a park dominated by a few weedy species. What the BNHS and the government of India wanted from Bharatpur, their idea of "natural," was in fact an avian-oriented ecosystem dependent on livestock grazing and human fodder and thatch collection. Bharatpur had never been a natural ecosystem—it had been created for duck hunting after all—and thinking that it would thrive without human management was simply wrong. If Bharatpur were to continue as one of the world's premier sites for avian biodiversity, it would have to be actively managed toward that goal.

The BNHS since 1991 has been asking the park managers from the forest department to please let the buffalo back into the park. Shruti Sharma, the forest officer in charge of Keoladeo Ghana National Park explained: "The BNHS had approached the government (specifically Sálim Ali) to push the establishment of Keoladeo as a national park. They wanted to end grazing and cutting in the park. After the park was established, and the BNHS conducted their ten-year study, they decided that the lack of grazing was hurting many elements of the avian habitat. They therefore began to advocate allowing cattle [both cows and buffalo] in the park. They have suggested that two to three thousand buffalo could be sustained in the park. Around my park I have forty thousand buffaloes. Whose should I allow in? And before I allow the buffalo in I have to change the Wildlife (Protection) Act."[43] As it is, there are feral and non-feral livestock in the park, which the forest department staff periodically

Figure 13. This cow is grazing in Keoladeo Ghana National Park in Bharatpur. Hundreds of cattle now live (illegally) within the park's boundaries and appear to play an important role in regulating vegetation growth.

Photograph by author, December 1997

tries to remove. It is not politically expedient to kill the cattle, so they are removed some distance from the park and released (where many eventually die of starvation anyway, and others are claimed by their owners). Many of the feral cattle in the park die after each monsoon of hoof-and-mouth disease, and the park staff have even tried sterilization of cattle in the park. Ultimately, Sharma admitted, it is a rather hopeless battle. Even if all the cattle currently in the park were removed, villagers periodically break holes in the walls and let cattle in.

Some of the local villagers who serve as naturalist guides at Bharatpur accept small bribes in order to let cattle come into the park, but even with imperfect enforcement, there are far fewer livestock in the park than there would be if the ban was lifted and far fewer than if the BNHS could manage the park according to their recommendations. Local forest officials like Sharma are in a bind—she admits that three or four hundred buffalo would be invaluable in clearing out canals of weeds. She is in a politically vulnerable position, though. If she began to hold lotteries to allow buffalo to come into the park, she would be attacked by locals for not allowing more buffalo into the park or for exhibiting bias in the selection process, and she would be attacked by environmentalists and government officials for breaking the

Wildlife (Preservation) Act, which had been such a battle, literally, to begin to enforce in the first place. As she said, until the law is changed, she will not change the park policy.

The forest department did make one compromise following the release of the BNHS study. On the basis of the unmanageable fires that had devastated much of its grasslands, the department decided to allow villagers to collect fodder in the park. Villagers pay a fee and get a cutting license and are then allowed to cut any grasses in the dry sections of the park. This policy has been in effect since 1991.[44] In the United States grassland management relies on frequent controlled burns that are managed so that they do not burn too hot or out of control (in New Mexico in the summer of 2000 this policy failed, as several fires did get out of control). To burn grass at Bharatpur would be a colossal waste of a resource for humans. This need to manage Bharatpur in such a way so that valuable forest resources are not wasted influences the options for clearing out the weeds as well. One U.S. scientist, Craig Davis, has suggested that water levels in the canals should be managed to mimic quick monsoonal floods, thereby drowning the weeds. Some BNHS staff members have suggested that perhaps more wild herbivores (nilgai and sambar) could be introduced to the park to eat more of the weeds—if they even *would* eat the weeds. Neither of these options has been judged particularly plausible by the forest department, and advocates for the local villagers argue that such complex solutions are not needed and would be wasteful compared to simply importing buffalo.

The BNHS Bharatpur study has never been published in a scientific journal. Copies of the final report have been bound and distributed widely throughout India, but informally. Even without publication, this study and its shocking determination that the banning of livestock hurt the health of the national park is one of the best-known research projects in India among ecologists and environmentalists. Many environmentalists accept the BNHS findings for Bharatpur but still maintain that larger parks that are not so artificial would benefit from a ban on grazing. It seems that more studies should be done—were the graduate students at Gir correct in their assessment? If so, that would support the implications of the Bharatpur study. Defenders of natural management in the United States argue that the management idea is good but that Yellowstone is not quite large enough, either.[45] Yellowstone is a huge park. Conversely, many pro-people advocates who work with villagers claim that the Bharatpur findings apply to grazing and fodder collection in all national parks. All

Indian ecosystems, they maintain, have evolved within the context of (sometimes, heavy) human use. This debate is being carried over into other national parks as researchers attempt to determine whether grazing is harmful or beneficial in different localities.[46]

The Bharatpur ecology study illustrated that some of the basic tenets of park management, as suggested by ecologists, do not always work. This policy was based on management principles developed in the United States for a national park system very different from Bharatpur yet in which the policy was similarly ineffective, though without the terrible human cost. The BNHS study showed that nature reserve management decisions often are predicated upon assumptions that have never been tested. These assumptions often are based on inadequately tested theories derived from ecological insights that seem so true that no further testing is needed. For generations of ecologists and park managers, in the United States and in India, the destructive nature of livestock grazing was so apparent that it never even needed to be discussed. At Bharatpur this received wisdom was wrong. The true tragedy is that this lesson was not free—its cost was paid in blood by villagers killed in the interests of a policy that ended up being harmful to the national park.

Even had the exclusion of humans and their livestock from the park been effective in maintaining the park's avian diversity, this case further illustrates the divide between the internationally connected and politically powerful ecologists and conservationists—those who made and advocated the laws—and the relatively powerless rural villagers surrounding the park. The exclusion of human influence from Bharatpur was advocated by the Indian scientists of the BNHS, was made possible by the Indian Wildlife (Protection) Act, and was enforced at the insistence of the Indira Gandhi–led Indian Board for Wildlife. The Indian National Park system corresponded to the U.S. model, with the preference for large areas, cleared of people, with no livestock, and with "natural management." Its application in India, though, was not due to an imposition of U.S. will on an Indian subject. Instead, the U.S. national park model corresponded to the conservation beliefs and values of an Indian urban conservationist elite (including Indian royalty, who used similar types of management policies for their hunting reserves, politicians such as Indira Gandhi, and scientists such as Ali) who then imposed this policy on rural Indians living in and around national parks and wildlife reserves. The irony of the futility of this imposition in actually preserving avian biodiversity makes even worse the reality of the violence done to rural Indians, and accepted, if not ordered, by Indian elites in the name of environmental preservation.

Preserving Nature in India

On July 20, 1980, India embarked on an attempt to establish the largest and most complex nature reserves imagined by ecologists to that time. On that date, the Indian National Man and the Biosphere Committee contacted Madhav Gadgil of the Indian Institute of Sciences and asked him to prepare a project document for a new type of nature park, a biosphere reserve, to be located in the Nilgiri Hills of southern India. The committee made it clear that political support for this project emanated from the prime minister's office. Gadgil immediately put his other research on hold and spent the next ninety days in the field under monsoonal conditions, surveying existing national parks and wildlife sanctuaries, as well as the areas around them. On October 21, Gadgil submitted his proposal for the Nilgiri Biosphere Reserve to the committee.[47] To that point the proposal was the most complex and sophisticated attempt anywhere to apply state-of-the-art ecological theory to the problem of preserving a specific natural ecosystem within a matrix of human settlement and use that had been written. As such, it drew heavily upon the guidelines of the UNESCO Man and the Biosphere program, guidelines that largely had been ignored by other nations. The Indian MAB committee met Gadgil's proposal with enthusiasm, and after some modifications the national committee began the long process of formal notification of the biosphere reserve. Finally, on September 1, 1986, with great fanfare and a national conference of scientists and forest officers, the Nilgiri Biosphere Reserve was formally established.[48]

By 1986, however, the Indian political landscape had changed dramatically, most notably with the assassination of Prime Minister Indira Gandhi in 1984, and the subsequent election of her less environmentally oriented son Rajiv as prime minister. Though the Nilgiri Biosphere Reserve was established, it was never formally joined to the UNESCO biosphere reserve network by the Indian government, nor was a working management plan ever approved. For all intents and purposes, the reserve was stillborn. In spite of this, thirteen other biosphere reserves have been "notified" in India, but not UNESCO, between 1986 and 1999, as the national MAB committee staggered along under its own momentum with the hope that someday these paper reserves would become real ones. Gadgil himself has largely abandoned the concept that he did so much to define, and is now widely perceived as an ardent supporter of small-scale, village-level biodiversity initiatives. Gadgil's story, and the story of Indian biosphere reserves, is useful in discussing the impact of the SLOSS debate (chapter 6) and associated questions about reserve design in India.

Biosphere Reserves

When Gadgil was contacted in 1980 to write the project document for India's first biosphere reserve, he knew exactly what he was being asked to do. Since his days as a graduate student at Harvard with E. O. Wilson in the late sixties, Gadgil had been aware of the development of the Man and the Biosphere program. The IBP ecologists had led the drive for a UNESCO conference in 1968 (the Biosphere Conference) that had first suggested biosphere reserves as new models for nature conservation.[49]

On the basis of the recommendations of the biologists at the Biosphere Conference, UNESCO, in a series of meetings in 1970 through 1972, established the Man and the Biosphere Program (MAB). MAB was centered around three ideas that were somewhat new within national conservation plans: (1) biodiversity must be saved within a matrix of *human* settlement,[50] (2) conservation should focus on entire ecosystems, not just single species, and (3) conservation should occur in a new model for national parks to be called biosphere reserves. The biosphere reserve portion of MAB, also called MAB Project 8, was the policy centerpiece of this initiative. The biosphere reserve was conceived as a "new" and "scientific" national park, one organized not around aesthetic considerations (as at Yosemite) or geothermal anomalies (as at Yellowstone), but around the findings of "modern" ecology. When Gadgil wrote his project document for the Nilgiri reserve, his second sentence thus read: "Modern ecological theory tells us that the purpose of conservation of biological diversity is best served by the preservation of a large, compact area." Gadgil then asked, "How large should biosphere reserves be . . . ? Ecological theory tells us that a biosphere reserve should be as large as possible, and as compactly organized as possible. . . . The rates of extinction of already existing species decrease as the area of a reserve increases, a result which has now been well substantiated by a careful study of the extinctions on islands of various sizes."[51] Gadgil understood that biosphere reserves were based on island biogeography, specifically the single-large position, and that that position had been written into international reserve design policy by the ecologists who had created MAB. The conferences in which this had taken place were remembered twenty-five years later as a time in which "scientific consensus" was reached regarding how best to preserve the environment.[52] The scientific consensus was possible

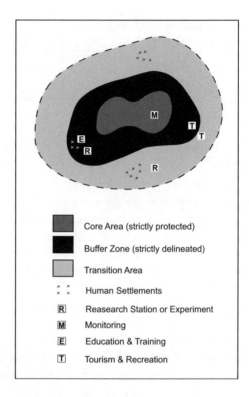

Core Area (strictly protected)

Buffer Zone (strictly delineated)

Transition Area

x x
x x   Human Settlements

R   Reasearch Station or Experiment

M   Monitoring

E   Education & Training

T   Tourism & Recreation

Figure 14.
Zonation of a biosphere reserve, courtesy
Orient Longman Private Limited

because advocates of the alternative several-small position were not present.

Biosphere reserves were not just large national parks, though. The area of a biosphere reserve was to be divided into different use areas, Thus, rather than thinking of a national park simply as an undisturbed core, surrounded by strip malls, or strip farming, a park could be conceived as a matrix of use areas: "a secure central area designed strictly for nature preservation, a surrounding buffer area in which relatively non-destructive multiple uses may occur, and an outermost transition area in which co-operative economic and research activities that are in harmony with the reserve occur."[53] Through regulating the buffer zone, the scientists who designed the biosphere reserves were hoping to control the extent of edge effect and more effectively preserve the pristine core of the reserve. Indeed, planning maps of biosphere reserves look almost identical to descriptive diagrams in ecology textbooks that show edge effect on a forest.

The new park architecture did not mean that the biologists of the MAB program were suggesting turning parts of existing national parks into buffers. Instead, they advocated the acquisition of more land surrounding existing nature reserves so that not just natural ecosystems might be managed but also all human activity surrounding the reserve. You might start with a national park as the inner core, for instance, but you would then expand outward, restoring degraded habitats and converting surrounding areas into limited-use buffers. As with Dan Janzen's Guanacaste National Park and Biosphere Reserve in Costa Rica, MAB suggests that the state should take control of even developed or degraded land—land under settlement or cultivation—and transform it into biodiversity refuges. Though the biosphere reserve ideal did acknowledge that humans live on or near nature reserves, rather than moving from that insight to an attempt to balance the needs of people and biodiversity, MAB suggested that people be moved out of the cores and into buffers, where their actions would be severely curtailed so as to be nonintrusive on the supposedly pristine core area. Not surprisingly, given the prevalence of conservation-oriented ecology in the United States, American ecologists were praised in the 1974 International Congress of Ecology by a MAB official as "being continually involved in the planning and development of MAB in general, and in particular MAB project area 8 [concerning biosphere reserves]." Following the initial planning meetings for MAB, U.S. ecologists had succeeded in generating "significant support in the U.S., both within the scientific community and in government."[54]

The United States and the Soviet Union were the first nations to establish biosphere reserves. In the best cold war fashion, when the Americans established three biosphere reserves, the Soviets established four. Both countries ignored some of the special elements of biosphere reserves—including their buffer zones and managed restoration areas. The United States simply added the biosphere reserve designation to preexisting national parks.[55] Park managers and ecologists were unwilling to convert any protected areas in national parks into buffer zones, and it was not feasible at the time to acquire additional land outside the parks.[56] Neither the United States nor the Soviet Union ever really intended to revamp their national park systems. Instead, it became clear that biosphere reserves were intended for the developing world. When Gadgil wrote his planning document for the Nilgiri Biosphere Reserve—complete with several core areas of hundreds of square kilometers, with buffer zones connecting the cores, with transition zones around the buffers, and with a few restoration zones for degraded habitats—he was doing something that had not been done before: he was taking the biosphere reserve concept seriously.[57] Since that time, other biosphere reserves have been established following the core-buffer model, but they have been sited in the developing world, often with the assistance of U.S. ecologists.[58]

At Prime Minister Gandhi's suggestion following the 1972 UN Conference on the Environment, India had put in place a national MAB committee. Established in 1974, this committee was responsible for funding studies on human-nature interactions throughout India. By 1979, however, at the urging of Indian scientists and some bureaucrats within the Ministry of Agriculture (there was no Ministry of Environment and Forests then), the decision was made to implement the biosphere reserve program in India. Experts were chosen to identify the country's biogeographic regions and a representative natural area within each region. This process (in which Gadgil played a small role) was completed in late 1979 and led to Gadgil's selection to write the first project document for the Nilgiri reserve. This reserve was intended to fulfill every requirement of a single-large advocate, encompassing 5,556 square kilometers spread across three states and including five national parks and wildlife sanctuaries as core areas connected by buffer corridors.

Between 1980 and 1986, Gadgil worked with forest officers and government officials to persuade them of the efficacy of his plan, including, after the assassination of Indira Gandhi, a completely new set of senior politicians. The government finally accepted his document, but with one primary change—they asked that he make the core areas of the proposed biosphere reserve smaller so that they would not include so many people who would subsequently have to be relocated.[59] The government bureaucracy was more concerned about the rights of rural villagers and tribals than the scientists were, in this case. Gadgil agreed to make these changes and in 1986 the Nilgiri Biosphere Reserve was formally notified in a large conference held at the resort hill station of Ooty, at the edge of the new reserve.

The notification of the Nilgiri reserve was bittersweet for Gadgil. Although his plan had been accepted, the new secretary to the newly created Ministry of the Environment and Forests,[60] T. N. Seshan, refused to allow India to join the International Biosphere Reserve Network of UNESCO. One of Gadgil's colleagues, a graduate student at the time, remembers him coming back from Delhi to Bangalore and announcing that the students were to stop working on the documents to register the biosphere reserve with UNESCO. Apparently Seshan gave no reason for his refusal, but this scientist suggested that Seshan felt India could do without international help.[61]

Despite formal notification by the government of India, without joining the international network of UNESCO the Nilgiri reserve was only a paper reserve. For it to work, it would have required coordination between several national and state government agencies, as well as the forest officers of the national parks and sanctuaries included in the biosphere reserve. This never happened, and no bureaucrat from the Ministry of Environment and Forests has ever encouraged it to happen. Without being connected to UNESCO, international funding for research and management did not become available—funding that Gadgil had relied on in his plans for managing the reserve. Low-level funding has been given by the government to various groups to conduct ecological research in the region over the last twelve years through the MAB committee, but otherwise, the Nilgiri Biosphere Reserve exists in name only, along with the other biosphere reserves throughout India that were subsequently notified but whose plans were never realized.[62] Had they been registered with UNESCO, UNESCO would have been invested in insuring that civil servants, both locally and in Delhi, took the program seriously. As the MAB representative from Paris explained at a 1998 conference in Delhi, UNESCO must wait for India to join the international network before taking any action.[63]

Between 1986 and 1990, Gadgil and the other scientists who had done so much in planning for the Nilgiri Biosphere Reserve held several conferences, trying to bring together the necessary forest officers and government agents to build a consensus in favor of MAB, but with little success. The local forestry officers seemed to have little interest in combining their individual national parks and sanctuaries (from three states) into a larger administrative unit. They also seemed uninterested in listening to ecologists. Only other scientists attended most of the meetings. This led to, or reinforced, the perception among some forester officers that MAB was a pet project of ecologists but irrelevant for managers. One retired forest officer blamed the failure of the reserves on the scientists: "They took all of the money for research, research on tigers and such, rather than [for] useful research for the people, and management."[64] He was mistaken in his details (the forest service has prevented Indian ecologists from working on tigers until recently, and the Indian ecologists at the CES who planned the park have never seen any of the research money). He was correct, though, in recognizing that the program was pushed by ecologists' concerns and activities and that at least initially, concern for local people was low on their agenda (as evidenced by the necessity of the government of India requesting that Gadgil revise his document to not include so many human communities in his core areas).

After speaking with dozens of scientists, forest officers, and bureaucrats with the Ministry of Environment and Forests—including three former secretaries to the Department (later, Ministry) of Environment and Forests—it has become clear to me that very few people know what really happened to the Indian biosphere reserve program. Almost no one in India thinks the concept was bad, and a variety of reasons have been proposed for why it did not work—some even mistakenly believe that UNESCO rejected India's biosphere reserves. According to Gadgil and Sukumar, Seshan had unilaterally decided not to join the international biosphere reserve network, but this is not widely known.[65] Only Seshan can give the definitive answer to why India's biosphere reserves were never internationally notified.

Since 1990, Gadgil has given up on biosphere reserves. He now believes that the program ultimately failed because "The central government personnel were interested only in pursuing [those things] which would give them more power."[66] They did not believe that biosphere reserves, linked in an international network and largely managed at the state level on the basis of the recommendations of scientists, would do this. This divide is a continuation of the earlier tension seen between ecologists, the IFS, and the government bureaucracy.

Indira Gandhi, strongly committed to India's position (and her reputation) as a leader in global environmentalism, might have made the difference for this particular project. Posthumously awarded the John C. Phillips medal, the highest honor of the International Union for the Conservation of Nature, Gandhi had been open to conservation.[67] A former secretary to the minister for the environment whose tenure overlapped with Indira Gandhi's assassination claimed that after her son Rajiv took office, "the bureaucrats got around him and told him this has to go [the staff of scientists heading the Ministry of Environment and Forests]. So, he brought a bureaucrat from Lucknow to head this, then another bureaucrat to succeed him."[68] The ministry is overwhelmingly staffed by members of the Indian Administrative Service, in stark contrast to when Indira Gandhi first established it and most of the staff were scientists personally selected by her. The biosphere reserve concept had the backing of every international environmental organization and the most well known ecologists in India and abroad, but it ultimately depended on the favor of the Indian government for success.

Small Reserves and Sacred Groves

Bharatpur, with its teeming birds, is the Indian crown jewel of any argument for small, managed biodiversity preserves that take into account human interactions within an ecosystem. Bharatpur is a completely non-natural environment. Its wetlands are the product of dams and dykes constructed by the maharaja of Bharatpur and on the other side of its brick wall are sixteen villages and thousands of people. Dozens of people everyday carefully remove dead wood for fuel (illegally) and grass for fodder (legally), and the park contains feral and tame cattle, grazing in the wetlands. Yet here you find, in twenty-eight nonnatural square kilometers, one of the most spectacular sites for avian biodiversity in the entire world. On the strength of this biodiversity it is a World Heritage Site.[69]

The World Heritage program, like biosphere reserves, is administered by UNESCO. Unlike the MAB (Man and the Biosphere) program, which was established by ecologists specifically for conservation, the World Heritage Convention was not written by ecologists, and its sites can be either natural or cultural wonders.[70] There are no ecological criteria for a natural area being declared a World Heritage Site—only the vaguer criterion of global uniqueness. Single-large ecological theory is not written into the requirements for the World Heritage Convention, as it is for biosphere reserves in the MAB program. Tiny Bharatpur's World Heritage Site nomination form could focus upon what it does have—one of the most phenomenal concentration of bird species of any place in the world.[71] In contrast, section 4.4 of the biosphere reserve notification form specifically requests proof that the proposed biosphere reserve "has an appropriate size to serve the three functions of biosphere reserves (this refers more particularly to (a) the surface area required to meet the *long term* conservation objectives of the core areas and the buffer zones)."[72] Bharatpur could never meet this criterion. Bharatpur also does not fulfill the biosphere reserve goal of preserving a functioning ecosystem. As demonstrated by the BNHS, this park needs intensive management, and its goal is not preserving a wide range of biodiversity, but preserving avian biodiversity. But could a series of small reserves like Bharatpur, intensively managed and oriented toward conserving different genera of flora and fauna, be India's several-small answer to the single-large approach of biosphere reserves?

For the last eight years, almost all of Gadgil's work has been oriented toward what he calls village-level biodiversity conservation. He has decided that if biodiversity is to be preserved in India, it will be because the people living in its midst have decided to save it. To this end, Gadgil is publishing local-language nature identification handbooks and training high school and college instructors from throughout southern India in techniques of biodiversity monitoring and preservation. In this time he has also coauthored two books of environmental history with Ram Guha, one of India's foremost environmental sociologists and an avid supporter of the several-small model for biodiversity preservation.[73] This is not a complete change of focus for Gadgil; he had long been interested in village-level biodiversity conservation, and ironically it was his work in that area that first brought him to the attention of the Indian national MAB committee, in the mid-seventies.

In 1971, when Gadgil returned to India (see chapter 5), he studied bandicoot rats, although he hated them. Soon thereafter Gadgil was rescued from the rats, and allowed to assist an old professor of his, V. D. Vartak, in a study of the sacred groves of Maharashtra. This was Gadgil's first conscious experience with sacred groves and he fell in love with these small forest fragments. He spent his days hiking through the hills and mountains of the Western Ghats, talking to local villagers, and surveying the biological diversity of their sacred groves. Gadgil worked on sacred groves for two years (1971–73) and in 1974 published the first of many papers he would write on this subject.[74] It was on the basis of his reputation for this work that Gadgil eventually became involved in the Indian MAB program, through the actions of S. P. Jain.

In his sacred grove articles, Gadgil argued that there already existed within India a sizable network of small biodiversity refuges, sacred groves, that should be officially included in any attempts at preserving India's biodiversity. He defined sacred groves as "tracts of sacred forests which have been completely or nearly completely immune from human interference on grounds of religious beliefs. . . . These sacred groves may range in size from a clump of trees to as much as 20 hectares in area."[75] Although the prevalence of sacred groves varies widely throughout India, they are particularly common in the Western Ghats and northeastern India, the regions of India commonly thought to possess the most biodiversity. Local villages maintain these sacred groves, and in some cases they represent undisturbed forests possibly hundreds of years old. Sacred groves within the states of Kerala and Maharashtra have yielded completely new plant species, and several other sacred groves contain the last remaining populations of certain plant species in that area. Although there have been hardly any entomological studies of the insects of the sacred groves, Gadgil suggested that such a study would also yield surprising results and may possibly uncover new species. Most important, the preservation of sacred groves would not be an imposition upon local communities that have been living with these small forests for decades, if not centuries.

Figure 15. Local hunters and herders traveling into Great Himalaya National Park leave offerings at this tree, thought to be a deity of the forests.
Photograph by author, April 1998

Sacred groves and the presence of human communities like the Bishnois in the state of Rajasthan (who are ideologically committed to carefully protect plants and animals living on their land) have become symbols for some Indian ecologists and environmentalists of an alternative model for biodiversity preservation, one that does not rely on large reserves and attendant dislocations of people or management models imported from the West. In this context, the several-small position has become linked not just with a particular set of ecological ideas but also with beliefs about the decentralization of government authority to local villages and to *panchayats* [traditional village councils]. Further, the several-small position has been linked to beliefs supporting the relative primacy of humans in relation to "nonhuman nature."

Ram Guha, a leading critic of large-scale reserves and what he calls authoritarian biology, has argued for the last fifteen years that ecological notions of unpeopled nature (i.e., wilderness) are an American cultural artifact and are inappropriate for anciently civilized and densely populated nations such as India. Conservation biology, with its attendant large nature reserves, was: "an ecologically updated version of the White Man's Burden, where the biologist (rather than the civil servant or military official) knows that it is in the native's true interest to abandon his home and hearth and leave the field and forest clear for the new rulers of his domain—not the animals he once co-existed with, but the biologists." In short, "National park management in much of the Third World is heavily imprinted by the American experience [and] the monumentalist belief that wilderness has to be 'big continuous wilderness' . . . leading to huge sanctuaries each covering thousands of square miles."[76] American ecologists, Guha argued, see ecology within a cultural framework that commits them to a specific set of ideas about wilderness and nature. How an ecologist perceives nature is central to how the actual science of ecology is conducted. In Guha's formulation, sacred groves and nonsacred, but small and human modified, parks like Bharatpur become the Indian solution to biodiversity reserves—small, local, indigenous—in contrast to large, globally networked, U.S.-imposed nature reserves. In short, several small reserves represent an appropriate solution for India.

For other Indian critics of Western science, such as Shiv Visvanathan, large reserves are symbols of the ecological version of modernization. For Visvanathan, science and development are inextricably linked and ultimately destructive for India (as well as for the rest of the developing world). Just as Western sciences such as physics and chemistry are held responsible for atomic blasts at Pokhran or gas disasters at Bhopal, so Western ecology has led India to create large nature reserves. Commenting specifically on biosphere reserves, Visvanathan suggests that "many a reserve . . . becomes the equivalent of the hydroelectric plant."[77] Visvanathan's critique of modernist science, with his emphasis on the dangers of large-scale projects indiscriminately applied throughout the world, leads him to advocate the formation of small-scale biodiversity refuges in which local villagers can develop appropriate management strategies—local science. Parks like the Ridge in Delhi—degraded, true, but still the haunt of mongooses and parakeets—or like Bharatpur or sacred groves are Visvanathan's best future for Indian nature conservation.

SLOSS, SLASS, or Neither?

The SLOSS debate—should India focus on large or small nature preservation models—drew me back to Delhi in 1998 to attend a conference expressly dedicated to the question of how to set priorities for national action for the preservation of biodiversity: the Biodiversity Conservation Prioritization Project. This conference was funded by a grant from the Biodiversity Support Program, a combination of U.S. funding agencies including WWF-USA, USAID, and the Nature Conservancy. The grant application had been prepared by Shekar Singh, a member of the Indian Institute of Public Administration who worked on environmental planning. This conference was the final step in a three-year effort to collect information and make recommendations about how to best preserve India's natural biodiversity. The program operated from the assumptions that hard decisions needed to be made and that scientific information would yield the needed data to help make those decisions. Thus, the conference included nearly every well-known ecologist from India and several from overseas. Thirty years after an international group of ecologists first suggested biosphere reserves at the UNESCO Biosphere Conference, another group of ecologists was meeting, but this time they were planning India's, not the world's, environmental future. And this time, there was no scientific consensus, false or otherwise, about how this should be done.

The SLOSS debate was recapitulated several times in the course of the conference, as participants argued about which strategy should be utilized. One speaker even reduced the two sides of the debate to two names, two Indian ecologists whose views and beliefs were generally understood to correspond to the respective positions—A. J. T. Johnsingh of WII for a single large reserve and Madhav Gadgil of CES for several small. Throughout the conference there were references to the Johnsingh or the Gadgil position, although neither researcher actually spoke to the issue. From being the author of the most ambitious large-reserve proposal yet seen in the world in 1980, by 1998 Madhav Gadgil had become a stand-in for the idea of a network of small reserves scattered across the landscape.

Gadgil's story seems pretty cut and dried, then. It is a story of a scientist trained in the United States who returns home and finds that the large reserves he learned to support while a graduate student at Harvard, the birthplace of island biogeography, just don't work in India, for any number of reasons. Therefore, he becomes an advocate for an alternative model of biodiversity preservation, one more in tune with Indian culture and demographics. That is the narrative that led people at the Biodiversity Conservation Prioritization Project conference to use Gadgil's name as shorthand for the several-small position, and the narrative that led many single-large ecologists I interviewed in India to claim that Gadgil was no longer an ecologist, but now merely a political advocate for the pro-people movement.

This is far too simple, however, and is a fundamental misreading of Gadgil's positions and beliefs. In his writings on village-level biodiversity conservation, Gadgil argues that small reserves *can* provide biological resources for people. Rural people who are used to living within forests and depend on forest products for their livelihoods cannot be denied access to biodiversity.[78] Small reserves also empower local people. Large reserves are still needed, though, for the preservation of biodiversity for posterity, and especially for the preservation of those species that could not continue in smaller reserves. Should SLOSS be replaced by SLASS (substituting "and" for "or")?

There is one other connection that needs to be made. In 1997, Gadgil was selected to be the chair of the scientific technical advisory panel for the Global Environment Facility (GEF)—the funding branch of the World Bank for environmental issues. This was a position with much power, but little recognition. As Gadgil points out, he is the first chair of that panel from the developing world. In that same year, the GEF announced that India had been awarded a $67 million grant for ecodevelopment. Shekar Singh, the organizer of the BCCP conference, had prepared this grant proposal with assistance from a small committee including Madhav Gadgil. According to Singh, "the only objective [of ecodevelopment] is to conserve the protected area [the national park], albeit in a socially just manner, and only that level of investment is legitimate which is required to divert unacceptable pressures from the Protected Area." [79] To that end, the ecodevelopment funds from GEF will be used to persuade people to move out of national parks and other core areas and then to live in a "sustainable fashion" around the protected area. This is, in effect, the biosphere reserve plan of managed buffer areas with controlled human habitation and large pristine cores, without the name but with one important difference—$67 million (over 2.6 billion rupees), will be backing up this plan. That is a lot of money, and it is all to be spent on a handful of large national parks—fifteen, currently. The single-large model lives on in India and with it, the fear that local people will experience an erosion of their autonomy, or even their livelihood, in the interest of creating large unpeopled parks.

Perhaps because of his position within the GEF, Gadgil did not speak about the ecodevelopment initiative. Still, I think it would be naïve to assume that there is no connection. It appears that Gadgil, now a senior ecologist, acted in the best tradition of the rain forest mafia. He moved into a position of power within a well-funded international organization dedicated to environmental conservation and assured that proposals he supported, even participated in preparing, received generous funding and support—and thus the literal landscape of Indian biodiversity preservation was changed to better reflect Gadgil's environmental values.

Gadgil's story is a richly suggestive one. The idea of a rain forest mafia seems quite different when Indians begin manipulating international environmental agencies to fund projects in India, as opposed to when Americans are doing so. And, after all, biosphere reserves have failed within India, and the ecodevelopment plan was written and conceived in India. It is true that this ecodevelopment plan drew upon many of the same assumptions as biosphere reserves, particularly with regard to pristine, peopleless core areas, but Gadgil also reflects in his other activities a clear commitment to the well-being of India's rural villagers and to a variety of different models of biodiversity preservation. In fact, many Indian readers of an earlier version of this text found it jarring to think of Gadgil as anything other than a supporter of rural peoples.

Gadgil has refused to accept the validity of the SLOSS debate or its underlying principle: that it is possible to find the conservationists' holy grail—a universal model of reserve design, applicable all over the globe, be it large or small. Instead, Gadgil's negotiations of the SLOSS debate suggest a scientific pragmatism in which he made decisions less on the basis of theoretical scientific unity than on the basis of effectiveness and possibility. He did not choose to work on sacred groves when he first returned to India, and he would not have chosen to abandon biosphere reserves in favor of ecodevelopment had biosphere reserves been an effective management option, but he was able to incorporate all these things into his work. Gadgil did not feel a need for theoretical consistency—he was and is quite content to advocate sacred groves and ecodevelopment plans simultaneously. Further, this was not a dialectical process—Gadgil has not tried to create a middle ground in the SLOSS debates. He rejects the idea that the divide is valid. Instead, he picks and chooses from the available options to create viable solutions for different problems. The entire SLOSS debate is, and was, misguided and counterproductive.

What this Indian history of biosphere reserves and of Bharatpur challenges is not just the validity of the Single Large position, or natural management, but the entire attempt to propose a universal scientific model of reserve design throughout the world. This is something so fraught with cultural values that it defies standardization. Given hard choices, what should be saved? How should it be saved? These are questions of value, of culture. An international model of reserve design assumes that there is one best way to save and preserve nature. Ecologists—scientists—look for universal, not just local truths. In so doing, they run the risk of imposing models developed for one area, and with one set of values in mind, upon another area and culture, which might then experience the model as a straitjacket. The best argument against international ecological expertise, or the attempt to impose universal conservation models, is that it has not worked. Biosphere reserves did not help India's nature. A peopleless, cattleless park did not help Bharatpur's birds. That is what we might learn from this history: the futility of imposing one vision of how to save nature across the globe, no matter how dire the crisis seems to be.

Conservation Ecology Crossing Borders

WHILE I WAS STAYING IN DELHI, I had the opportunity to attend a concert by L. Subramaniam. Subramaniam plays the violin, but he plays it in the Indian style, sitting cross-legged on the floor with the violin held toward his legs. The violin, though an instrument originating in the West, has been used in Carnatic (classical Indian) music for centuries now. The violin's four strings are usually set to different notes by Indian violinists—rather than the European violin's E-A-D-G tuning, an Indian violinist might tune the instrument to E-A-E-A, or D-G-D-G. The violin is fretless; there are no bars on the neck of the instrument that divide the strings into exact musical intervals corresponding to the chromatic, Western, scale. This means that the violinist, by moving his or her fingers a slight distance up or down the string, can easily bend and modulate notes by microtones, intervals between notes that a fretted Western instrument (like a guitar) could not play. Classical Indian music cannot be played without these microtones.

In the concert that I attended, Subramaniam performed with other musicians who played a mixture of traditional Indian and European instruments.

Subramaniam writes much of the music that he plays—one piece that he played in his concert that night was a concerto for two violins. In this piece, one violinist began by playing in the style of the European classical tradition, Subramaniam, the second violinist, then played in the Indian classical style. As the piece progressed, the style of each violinist gradually moved closer and closer to that of the other, until they finally played together in a synthesis of these two musical traditions. It was stunning.

Subramaniam's concert was a celebration held in conjunction with the fiftieth year of Indian independence. That same month an Indian friend, Dhiraj, took me to the dress rehearsal for the annual "Beating the Retreat," a ceremony commemorating the end of the Republic Day festivities held each January in honor of India's independence from Britain. This ceremony is centered on a series of marching band performances. As the sun set, the bands marched between the stately buildings built by the British in the 1930s for their capital, now the governmental buildings of independent India. Military sentries, including men on camels and horses, stood silhouetted against the sky. As the bands marched, they played such Christian standards as "Abide with Me," and "Amazing Grace," complete with bagpipes.

The evening had a tinge of the bizarre to it. There was pomp and pageantry, certainly, and without a doubt I was glad to see it. But the whole thing seemed a bit incongruous, a relic of the British colonial state that had hung on for half a century, now. The band did not play only Christian hymns—it also played pieces by Indian composers and, of course, the Indian national anthem. Even these pieces sounded odd. Perhaps cultural syncretism does not always work?

Marching bands are made up of instruments that, like the trumpet, can play only the chromatic Western scale. A trumpet can change the notes it plays by using three valves, which allow the air to pass through different lengths of the metal tubing that make up the instrument. The valves, in combination, allow the trumpet to play the complete Western chromatic scale. A trumpet, however, cannot play microtones. Sure, you can try, by pushing the valves partway down or by bending the pitch with your lips. But this is a limited solution at best. A trumpet, in short, cannot play Indian classical music or even Indian folk music—at least not in a style that sounds somewhat Indian. And, by extension, a marching band organized around the trumpet and its brass kin will never play an Indian raga. Some things—when they cross cultural or geographical borders—are more or less malleable than others, and this is true of science, ecology, and conservation practices, just as it is true of musical instruments. Western musical forms and instruments have spread into India. Subramaniam and his violin are a testament to the power of cross-cultural exchange and mixing to create something new, something beautiful: creative innovation. Other times, as with an Indian marching band playing "Amazing Grace," the mixture of cultures sounds to me decidedly less like a mixture and more like the spread of an art form, largely unchanged, to another region of the world. It can still be beautiful to the eye and ear, but its Western elements prevail. Cultural imperialism assumes that all global exchanges are trumpets—fairly inflexible products of one culture, which, as soon as you pick one up, destine you always to play the same types of tunes. Globalization suggests, to the contrary, that all exchanges are violins, infinitely malleable to local interests and traditions. Diffusionists would not care, because everybody wants to play Beethoven and Sousa anyway.

When conservation-oriented ecology crosses national borders, is it a violin or a trumpet, cultural imperialism or globalization? Both seem appropriate, depending on the example chosen. When we consider Sukumar's use of evolutionary ecology to investigate elephant-human interactions, ecology seems like globalization. Likewise, when the Harvard-trained Gadgil studied the sacred groves surrounding Pune, this science seems flexible. In these cases, uniquely Indian concerns were combined with ecological theories created in the United States to create a new type of ecological study.

At other times, conservation-oriented ecology has seemed much more like cultural imperialism, as with the unconsidered application of island biogeography in keeping with SLOSS orthodoxy, or when American and Indian ecologists advocated the removal of all cattle and human uses in Bharatpur. Many scientists would seek to disallow these examples as science at all—but this is how science is practiced, at its worst; this is science in a straitjacket. Then there is a vast middle ground of studies that seem not to fit perfectly in either category, and the metaphor begins to break down.

When Sálim Ali persuaded Dillon Ripley to collaborate with him to create the *Handbook to the Birds of India and Pakistan,* how should we understand this? Ali and Ripley did not create a new theory. Ripley did not change established ways of ordering birds in taxonomies; Ali did not create a new science of bird behavior. Ali was placing India's birds into a Western schema, adding to the edifice of Western science. A cultural imperialist might claim that Ali was a supremely skilled trumpet player—better than most of the trumpet players in the West, but a trumpet player nonetheless. And with Ripley, he could play quite a duet. But Ali was reclaiming knowledge about Indian nature from colonialism—he was an

Indian cataloguing India's nature. He went to Germany and brought back a new type of ecology to India—for Ali, taxonomy was the trumpet that allowed no flexibility. The behavioral ecology idea, in contrast, was liberating for Ali. It was developing local knowledge. It allowed him to do something, learn things, that had never been done before—to make new music, as it were. And after all, violins were invented in Europe—it is not the origin that matters but how the instrument is used.

It would be easy to fall into the trap of attempting to fit the different people and events of this history into neat explanatory categories—globalization or cultural imperialism (or, metaphorically, trumpets or violins). To do so, though, would be to fall into a trap very similar to the one I just criticized in the previous chapter—the SLOSS trap of believing that you can separate two alternatives to a question and assume that one is correct and the other wrong. Throughout the development of ecology in India in the last fifty years, several stories and trends defy easy categorization. Globalization and cultural imperialism, after all, are meant to be models to help us understand how human history occurs. Both offer us some insight, but, at least for ecology in India and the United States, accuracy lies in some combination of different elements of these ideas.

The idea of cultural imperialism implies that things have pure origins. This is a simplistic assumption. What was the pure origin of the evolutionary-ecology synthesis? German behavioral ecology? American community ecology? Darwin? The interactions between European and non-European people and lands during the era of colonialism? The origins of ideas and schools are not pure. Ripley learned ecology from Hutchinson, an Englishman, and Mayr, a German, in the United States, while Ali practiced German ecology in the heart of the British Empire. Who introduced ecology into India? Ali? Stresemann? Gadgil? Schaller? E. O. Wilson? By now you know that the story is much more complex than this. Scientists from throughout the world—not just the United States—developed the science of ecology. And just as Americans were concerned with vanishing wilderness, Ali and the BNHS were concerned with the loss of wildlife in India, and the British were concerned with environmental destruction in their empire. The roots of conservation go back much further than the mid-twentieth century, or even the formation of Yellowstone National Park in 1872. Although the U.S. ecologists, and their theories, played a disproportionate role in global conservation in the latter half of the twentieth century, they did not play the only role. I traveled through the journals, debates, forests, and institutions of India with a very specific pair of spectacles on, looking for links between American and Indian ecology. I am open to the possibility that in so doing I may have understated the importance of other links, trends, or developments.

Further, cultural imperialism often assumes that cultures correspond to state boundaries. When India's cattle were criticized, an Indian conservation elite led the charge, with the support of U.S. scientists. I have argued throughout this book that often, Indian scientists, conservationists, or government leaders have more in common with U.S. ecologists than they do with India's rural poor (and their policies and conservation advocacy sometimes reflect this). They speak English. They fly around the world to conferences and workshops. They live in a technologically complex environment. They sometimes repudiate caste, or other "traditional" social practices, such as the wearing of traditional clothing. It seems that many Indian biologists belong to a transnational "comfortable" class, sharing more with their fellow American scientists than with their poorer compatriots. There has been a lack of sympathy at times among many Indian ecologists and conservationists for rural Indians, in the aggregate, living in and around (potential or actual) protected areas. This was seen most clearly at Bharatpur, but it was also there when Gadgil's proposed Nilgiri Biosphere Reserve had core areas with so many people inside them that the government of India asked him to redraw the boundaries so that fewer people would have to be relocated. This is especially surprising considering Gadgil's later stellar efforts in human-centered ecology.

In spite of its flaws, cultural imperialism does force us to confront the exercise of power—and that clearly matters. Throughout this history, there are numerous examples of the importance of state politics, of wealth and social status at birth, or access to political and institutional power. The United States has benefited greatly from its economic power. The PL 480 fund, which arose from U.S. foreign aid (itself a product of U.S. prosperity), enabled a great many American ecologists to work in India. Lesser grants had a similar but more localized impact. One of the intriguing lessons of this history, though, is that the government of India did have the power to limit U.S. influence—and in fact, when the government attempted to shut down U.S. projects, they were successful. This was not necessarily good for Project Tiger, but it did demonstrate that India was not a lackey to the U.S. state. The Indian state's power to control its territory trumped America's economic power to bribe its way into India's national parks, at least for a few years.

The most compelling power exerted by the United States in this history was not political, military, or economic. It was the power of some of the U.S. ecological ideas—particularly the development of the evolutionary-ecology synthesis that, while its origins can be traced throughout the world, truly flowered in the United States. Such things occur from time to time throughout the world—communities of genius arise, then (usually) quickly pass. It is as if a certain combination of cultural factors and a synergy of similarly committed individuals result in some unexpected outpouring of brilliance. This is the story of the German physicists in the early twentieth century, of Italian artists during the Renaissance, or the activists of the Indian independence movement. And it seems to have been the story of the unusual confluence at Yale and Harvard of the British Hutchinson, the German Mayr, and their American colleagues and students—MacArthur, Wilson, Ripley, Slobodnick, the Odem brothers, Simberloff . . . giants all, in their way. This is a strange type of power, though—more globalist than imperialist.

We began by considering Christy Williams and his research on elephants in Rajaji, and I suggested that a richer understanding of the history of conservation science in India, and its links with the United States, was needed before we could attempt to understand his research. So now, what do we make of his study? The links outward from Williams touch every corner of this book's story. He was a graduate student at an Indian university. He did not need to leave India to do his Ph.D.; he could pursue his graduate study at the WII, supported by the government of India. Before his dissertation research he had been sent across India by the WII to prepare reports on different species and different reserves, and even an environmental impact assessment. He was studying with a large USFWS grant—a legacy of the PL 480 shift away from the Smithsonian and to the USFWS in 1979, in part due to the Smithsonian's army taint. As part of this grant, he had access to U.S. equipment, and he had a U.S. collaborator. His U.S. collaborator was not doing the research himself but only visited the field site, and made it possible for Williams to come to the United States to do work on his data. Williams's advisor was A. J. T. Johnsingh, recruited by the BNHS to be the field assistant for Michael Fox. Johnsingh was the first Indian to get a Ph.D. on an Indian mammal study, and he now heads the biological faculty of the WII. Johnsingh relied, while getting his Ph.D., on the help and advice of the young Madhav Gadgil at CTS, newly returned from Harvard, where he had picked up the latest in ecological theories and techniques. Johnsingh learned radio telemetry at the Smithsonian Institution's field station in Virginia, a continuing legacy of Ripley's interest in promoting the ecological sciences through that institution. Williams was the first WII researcher to use radio telemetry on elephants—the legacy of the government of India's suspicions regarding that technology. He was conducting the latest elephant study in India—following in the footsteps of Gadgil's former student Sukumar. Williams was working in a national park that has human use and cattle. His study helped provide the information on the park's ecology that will help to determine what will happen to those people. Following Bharatpur, herders are not forced out of protected areas at gunpoint.

Have U.S. concerns, agendas, actions, or theories shaped Williams's study? Clearly. But so have Indian concerns, agendas, actions, and theories. It is difficult to imagine how Indian ecology would have developed in the absence of the United States, so intertwined are the two nations' ecological sciences, but at the same time it is clear that Indian ecology developed along its own lines, fulfilling the goals of Indian actors, be they scientists, activists, or bureaucrats.

Acknowledgments

I COULD NOT HAVE COMPLETED THIS PROJECT without the generosity (of spirit, ideas, and funds) of numerous individuals and organizations in both India and the United States. A complete list would be impossible, but I would like to recognize a few key figures and institutions.

My parents were my first teachers and I thank them both for the love of learning they instilled in me. I have been fortunate to work with several mentors through my years at the University of Iowa, the University of Alabama, and Rhodes College. An earlier version of this book was my dissertation. While working on this project at Iowa I had two superb advisors and friends, Paul Greenough and Rich Horwitz, who made my project and graduate student career not a hurdle but a joy. Without their encouragement and help I would never have undertaken such a project. I also thank all the members of the American Studies community at Iowa— my friends and colleagues who shared their ideas, their time, and their love of life. Thanks also to Lynn Adrian, Ken Cmiel, Jim Lanier, Susan Lawrence, and John Nelson for their work with me. I have been fortunate in finding another good community of colleagues and friends at Salisbury University.

Several individuals have provided me with extensive comments on my manuscript. This book would have been far poorer without their assistance—thanks particularly to Ram Guha, Mahesh Rangarjan, and Vasant Saberwal. Needless to say, they are not to be held responsible for the book's errors of fact or judgment.

Much of my dissertation involved research in India. That research was made possible by the help of several key figures. Without them this project would not have been completed and certainly would not have been nearly as enjoyable. Ravi Chellam, Raghavendra Gadagkar, Shiv Visvanathan, and Subhash Chawla insured that my time in India was intellectually productive and physically possible. I met with countless scientists, administrators, and activists throughout India, and I thank them all. I am particularly indebted to the U.S. Educational Foundation in India, the Wildlife Institute of India, the Bombay Natural History Society, the Centre for Ecological Sciences, WWF-India, and the Centre for the Study of Developing Societies for offering me access to their libraries, their staff, and their expertise.

Not to be forgotten, graduate students throughout India took me with them to the field, shared with me their ideas, their food, their good times, their soccer games, and their expertise. At times, as when I was panting up the Himalayas, or noisily slopping through the rain forest, I am sure I disrupted their research, but never did any of my gracious graduate student hosts ever make me feel anything other than completely welcome. More than anyone else, they bore the burden of my research, and they bore it with grace. My thanks to them all.

The Smithsonian Institution Archives were an essential help to my project. Thanks to the dedicated archivists on staff there, and especially Pam Henson, Bill Cox, and Tracy Smith. The family of S. Dillon Ripley generously granted me permission to access their father's papers—stored, but restricted until 2011—at the Smithsonian Institution Archives. My work on the U.S. side of the Indian-U.S. exchange would have been seriously limited without this access and I offer to them my sincere gratitude.

I thank the Fulbright Foundation and the U.S. Educational Foundation in India; the T. Anne Cleary Foundation; the Seashore Fellowship program, administered by the Graduate College of the University of Iowa; the Charles R. and Martha N. Fulton School of Liberal Arts Foundation, Salisbury University; the Alexander Kern Travel Fund; and the American studies program of the University of Iowa.

I first encountered biosphere reserves when reading a paper by a good friend—that same friend later introduced me to island biogeography and metapopulations in the prairies of Iowa, to extinction debates, and to international travel. Unseen in these pages, but lurking behind every paragraph and present at every field site and bumpy bus ride, is my favorite informant on the world of the ecological sciences, my personal field guide, Cassy Kasun Lewis.

Notes

Abbreviations

acc.	accession number
CD	collection division
f.	folder
int.	interview
JBNHS	*Journal of the Bombay Natural History Society*
RU	record unit
SIA	Smithsonian Institution Archives

Unless otherwise indicated, all interviews were conducted by me.

Introduction

1. In Indian-English, Williams would be considered a postgraduate student. I have chosen, however, to use the American-English term *graduate student* to refer to anybody working on their master's or Ph.D.

2. WII, *Annual Report, 1986–87* (Dehradun: Wildlife Institute of India, 1987), 14.

3. Ullas Karanth, who studies tigers in Nagerhole National Park in the south of India as part of an Indo-U.S. collaborative project, was caught in the middle of a furious storm of criticism when five tigers in the park died within two months in 1990. One had been radio collared and died of old age. Despite a lack of connection between the deaths of the tigers and collars, politicians and the press widely called for an end to radio tracking of tigers in the park, and the research was temporarily halted. Raman Sukumar discusses this case in "The Nagerhole Tiger Controversy," *Current Science* 59, no. 23 (December 10, 1990): 1213–16. A more contemporary case concerns a colleague of Williams's working on sea turtles, whose permission from the government of India was so delayed that the radio transmitters he was to use (whose batteries had a limited life) had to be sent on to Mexico instead. "Turtle Project," *Science* 276 (1997): 1785.

4. Raman Sukumar, *Elephant Days and Nights: Ten Years with the Indian Elephant* (New Delhi: Oxford University Press, 1994).

5. Shekhar Singh, Indian Institute of Public Administration, pers. comm., March 18, 1998.

6. All information on the Gujjars is derived from conversations with the Gujjars,

translated by Williams, and from P. S. Poti, *Rajaji: The Indian People's Tribunal on Environment and Human Rights: Preliminary Report on the Rajaji National Park* (Bombay: Indian People's Tribunal, 1994).

7. Interview with a retired IAS officer who has chosen to remain anonymous, WWF-India Offices, July 14, 1998. He claimed that his relocation of villagers at Kanha when he was the local Indian Administrative Service officer was the first such relocation attempt in India. He realized both the ease with which relocation could be done (by offering attractive resettlement options) and the positive effect it had on the park, even in the space of a year.

8. See, for instance, Mark Spence, *Dispossessing the Wilderness: Indian Removal and the Making of the National Parks* (New York: Oxford University Press, 1999).

9. Although many tribal communities are practicing settled agriculture at the encouragement of the national government and many male tribal people have migrated to larger cities to find wage labor, there are still several tribal communities living within India's forests as slash-and-burn agriculturists or as hunter-gatherers. P. Sainath, *Everybody Loves a Good Drought: Stories from India's Poorest Districts* (New Delhi: Penguin, 1996).

10. Ashish Kothari et al., *Management of National Parks and Sanctuaries in India: A Status Report* (New Delhi: Indian Institute of Public Administration with the Ministry of Environment and Forests, 1989), 79. Two-thirds of India's sanctuaries and national parks were reported as having unresolved land use conflicts, whether for people still living in the park or for those who were using the sanctuaries as a resource for grazing, fodder, and so forth.

11. Among the more prominent NGOs are the Rajaji Park for People support group (Ghad Kshetra Mazdoor Sangharsh Samiti) and Vikalp. From Poti, *Indian People's Tribunal,* 5.

12. Ibid., 7.

13. Ibid., back cover.

14. H. S. Panwar and B. K. Mishra, "Rajaji National Park: Real Issues, Problems, and Prospects," *Himachal Times,* October 6, 1994, 7.

15. A. J. T. Johnsingh and Justus Joshua, "Conserving Rajaji and Corbett National Parks: The Elephant as a Flagship Species," *Oryx* 28, no. 2 (April 1994): 135–40.

16. A. J. T. Johnsingh, "Rajaji," *Sanctuary* 11 (1991): 14–25; Panwar and Mishra, "Rajaji National Park," 6; Poti, *Indian People's Tribunal,* 6.

Chapter 1

1. In interviews with the faculty of the Indian Institute of Sciences, the Wildlife Institute of India, and the Bombay Natural History Society, Schaller's name invariably came up as the pivotal early figure in Indian ecology, followed by Americans Mel Sunquist (who worked on tigers in Nepal), Michael Fox, and Steve Berwick, as well as key Indian figures like Sálim Ali, Madhav Gadgil, and A. J. T. Johnsingh.

2. Including Ullas Karanth, who eventually attended the University of Florida, and Renee Borges, who attended the University of Miami.

3. Again, this is derived from my interview data, where one of my standard questions asked for a short list of preferred journals, both for reading and for publishing. Of course, just because a journal was listed as important does not mean that it was regularly read. One of my informants, for instance, regularly looked at *American Naturalist* but admitted that its statistics baffled him, and he seldom read the actual articles.

4. For an early classic statement of this principle, see George Basella, "The Spread of Western Science," *Science* 156 (1967): 611–22.

5. David Wade Chambers, "Does Distance Tyrannize Science?" in *International Science and National Scientific Identity,* ed. R. W. Home and S. G. Kohlstedt (Boston: Kluwer Academic Publishers, 1991), 32.

6. There are many examples of scholarship that suggests the importance of colonial science and scientists to the imperial center. A notable example dealing with India is Deepak Kumar, *Science and the Raj, 1857–1905* (Oxford: Oxford University Press, 1995).

7. Richard Grove, *Green Imperialism: Colonial Expansion, Tropical Island Edens, and the Origins of Environmentalism, 1600–1860* (New York: Cambridge University Press, 1995).

8. R. W. Home, introduction to Home and Kohlstedt, *International Science,* 12.

9. Chambers, "Does Distance Tyrannize?" 32.

10. Peter Golding and Phil Harris, eds., *Beyond Cultural Imperialism: Globalization, Communication, and the New International Order* (Thousand Oaks, Calif.: Sage Press, 1997), 4.

11. Bruno Latour and Barry Barnes were particularly influential in this development within the sociology of science.

12. Ramachandra Guha, "Radical American Environmentalism and Wilderness Preservation: A Third World Critique," *Environmental Ethics* 11, no. 1 (Spring 1989): 71–83.

13. Guha argues in several articles that, although ecology is practiced throughout the globe, U.S. scientists have provided many of the dominant ideas, theories, and experiments in the last half century.

14. Shiv Visvanathan, *A Carnival for Science* (New Delhi: Oxford University Press, 1996), 6.

15. This argument is similar to the one made by feminist critics of Western science, such as Carolyn Merchant, who see in all of Western science a unifying (destructive) worldview based on European cultural attitudes toward nature that permeates the practices and theories of that form of epistemology.

16. For a further analysis of Visvanathan's argument see Michael Lewis, review of *A Carnival for Science* by Shiv Visvanathan, *Iowa Journal of Cultural Studies* 18 (1999): 119–21.

17. This tendency is well documented in John Tomlinson, *Cultural Imperialism* (Baltimore: Johns Hopkins University Press, 1991).

18. Golding and Harris, *Beyond Cultural Imperialism,* 4.

246 | Notes to pages 21–28

19. Malcolm Waters, *Globalization* (New York: Routledge, 1995).

20. Arjun Appadurai, *Modernity at Large: Cultural Dimensions of Globalization* (Minneapolis: University of Minnesota Press, 1996), 31; Paul Hoch, "Whose Scientific Internationalism?" (review), *British Journal for the History of Science* 27 (September 1994): 349.

21. It also led me to oral histories with current and former ecologists, environmental activists, and governmental officials, as well as work in archives at several different NGOs, scientific institutions, and government offices. For a more complete description of my research methodology and sources, see my note on sources in the bibliography.

Chapter 2

1. The BNHS has refused to change its name to the Mumbai Natural History Society, deciding that it had too much invested in its original name. I refer to the city as Mumbai in the present, but Bombay when discussing the past.

2. R. E. Hawkins, ed., *Encyclopedia of Indian Natural History: Centenary Publication of the Bombay Natural History Society, 1883–1983* (Delhi: Oxford University Press, 1986), ix, 73. Also, S. Dillon Ripley, "The View from the Castle," *Smithsonian* 14 (December 1983): 10; J. C. Daniel, ed., *A Century of Natural History* (Bombay: Bombay Natural History Society, 1983). Further general information on the BNHS was gleaned from interviews with the staff of the BNHS in February 1998.

3. For multiple examples, see Daniel, *Century of Natural History.*

4. Sálim Ali, *The Fall of a Sparrow* (Delhi: Oxford University Press, 1985).

5. Ali's numerous prizes include the John C. Phillips Memorial Medal, the highest honor of the International Union for the Conservation of Nature (IUCN), the second ever J. Paul Getty International Prize for Wildlife Conservation, as well as national awards from the United States, Britain, and the Netherlands, and the Padma Vibhushan and Padma Bhushan from the president of India.

6. Madhav Gadgil, "Sálim Ali (12 November 1896–20 June 1987)," *Current Science* 71, no. 9 (November 10, 1996): 686. Similar sentiments were expressed by E. P. Gee, *The Wildlife of India* (London: Oxford University Press, 1964).

7. My investigation of Ali's life is based on his autobiography (written in the last years of his life), on interviews with colleagues, on his scientific publications, on the correspondence between him and his friend and collaborator S. Dillon Ripley (at the SIA), and on several shorter (usually hagiographic) accounts of Ali published by his students, friends, and colleagues.

8. Ali, *Fall of a Sparrow*, 3. The author of an autobiography can engage in exactly this sort of hedging—Ali, like many conservationists, began as a hunter. As with most hunters, he certainly thrilled in the capture of the rare and unusual more than the mundane house sparrow. Contemporary conservationists would not see that as a virtue, however, and Ali was writing this autobiography for an audience who recognized him as the father of Indian environmentalism.

9. S. H. Prater was one of many Indian-born Englishmen. He became curator of the BNHS for nearly twenty-five years before leaving India when it gained independence. Hawkins, *Encyclopedia,* 73.

10. B. G. Deshmukh, foreword to *Sálim Ali's India,* ed. Ashok Kothari and B. F. Chhapgar (Delhi: Oxford University Press, 1996), 10.

11. Ali, *Fall of a Sparrow,* 20; J. C. Daniel, "Unforgettable Sálim Ali," in *Sálim Ali's India,* ed. Ashok Kothari and B. F. Chhapgar (Delhi: Oxford University Press, 1996), 17.

12. Daniel, "Unforgettable Sálim Ali," 16.

13. This is a standard theme in histories of the Indian independence movement. Gandhi and Nehru are often caricatured as representative of two different visions for India's future brought together by their shared desire for Indian independence. In this dyad, Nehru is labeled an unabashed supporter of Western science, of big dams, and of modernization, while Gandhi is portrayed as an advocate of decentralized, village-level technology and non-Western modes of development.

14. Ali, *Fall of a Sparrow,* 30–32.

15. Daniel, "Unforgettable Sálim Ali," 16.

16. Hawkins, *Encyclopedia,* 73. Of course, Ali's family connections with the BNHS (in addition to his friendship with Prater) were crucial in his getting this job with the BNHS. A young Indian man without a college degree and without family connections would have had virtually no chance of being hired by the BNHS or of having his biological aspirations taken seriously.

17. Gadgil, "Sálim Ali, Naturalist Extraordinaire: A Historical Perspective," *Journal of the Bombay Natural History Society* 75 (supplement, 1980): iv.

18. For more information on these early naturalists in India, see Kumar, *Science and the Raj,* or Grove, *Green Imperialism.*

19. British botany had a much longer history in India than zoological studies, with the establishment of the Indian Botanic Gardens in 1787. Hawkins, *Encyclopedia,* 73–74. Interest in botany was often spurred by agricultural priorities: What European, African, or Western hemisphere crops could grow successfully in India? Similarly, what indigenous Indian crops could be grown for profit in Western markets? Most early interest in zoology was instead spurred by hunting—there was no easily identified economic benefit to be derived from animals. This held true not just in India but throughout the world, as botany was typically better funded and more sophisticated than animal studies. Entomology was in many places the most advanced zoological branch, because insect pests affected agricultural crops (thus many early animal ecologists were entomologists). See Paulo Palladino, *Entomology, Ecology, and Agriculture: The Making of Scientific Careers in North America, 1885–1985* (Amsterdam: Harwood Academic Press, 1996).

20. There is a great deal of debate about what counts as a typical specimen for any bird, as a certain amount of variation is natural. In a tight debate ornithologists rely on "type" specimens, stuffed birds that are the equivalent of the official platinum measurement standards (meter, kilogram) kept in Paris. Since there is usually more

than one type specimen, including at least one male and one female, a certain amount of variation becomes codified as natural.

21. Interview with Saraswathy Unnithan, Bombay Natural History Society, February 12, 1998.

22. Sálim Ali, *The Book of Indian Birds,* 12th ed., rev. and enl. (Delhi: Oxford University Press, 1997 [1941]), xxiii, emphasis in original.

23. Sálim Ali, "The Moghul Emperors of India as Naturalists and Sportsmen, Parts 1–3," *BNHS* 31–32 (1927–28).

24. See Sharon Kingsland, *Modeling Nature: Episodes in the History of Population Biology,* 2d ed. (Chicago: University of Chicago Press, 1995).

25. Ibid., 11.

26. Frank Golley, *A History of the Ecosystem Concept: More than the Sum of the Parts* (New Haven: Yale University Press, 1993), 17.

27. The move to quantification was actually quite common at the turn of the century. As outlined in Theodore Porter, *Trust in Numbers: The Pursuit of Objectivity in Science and Public Life* (Princeton: Princeton University Press, 1995), quantification was a tool for professionalization and increased status for many different professions. Within ecology it helped to differentiate scientists from amateur naturalists, and it gave the ecologists validity in the eyes of other scientists.

28. Ali, *Fall of a Sparrow,* 57.

29. What Ali interpreted as an anti-India bias was perhaps part of a larger antidominion bias. Kingsland records that when Australian scientists tried to come to Britain for training, they were denied. British scientists claimed "the presence of temporary workers from the Dominions would prevent the building of a vigorous research institution." Kingsland, *Modeling Nature,* 135. Kingsland is writing of the 1930s, and Ali had tried to go to Britain in 1929. It seems possible that British scientists, besides their concern about "vigorous" research, were also concerned about funding scientists from outside Britain. Kingsland does write that funding was scarce and that the British Bureau of Animal Populations required funds from the New York Zoological Society in order to run. Kingsland, *Modeling Nature,* 132.

30. Oskar Heinroth later began to bridge the gap between the study of bird behavior and taxonomy (by including behaviors as distinguishing characteristics of species). Ernst Mayr emigrated to the United States shortly after Ali left Stresemann's lab and, at Harvard, became arguably the world's most influential ecologist in the forties and fifties. If the ecology idea had been the first shift within the study of organismal biology, Mayr would help spearhead the second shift in the 1940s—the understanding of communities and individual behavior as the product of natural selection, a combination of genetics and evolution called the neo-Darwinian synthesis by some or, more grandly, the modern synthesis. The first term is quite common, as evidenced by its usage in even popular sources such as Paul Ehrlich, David Dobkin, and Darryl Wheye, *The Birder's Handbook* (New York: Simon and Schuster, 1988), whereas the second is less common: see, for example, E. O. Wilson, *Naturalist* (Washington, D.C.: Island Press, 1994).

31. Kingsland, *Modeling Nature,* 53–56.

32. Ali, *Fall of a Sparrow,* 57.

33. Zafar Futehally, "A Portrait of Sálim Ali," in *A Bundle of Feathers: Proffered to Sálim Ali for His Seventy-Fifth Birthday in 1971,* ed. S. Dillon Ripley (Delhi: Oxford University Press, 1978), 5.

34. Daniel, "Unforgettable Sálim Ali," 16; Ali, *Fall of a Sparrow,* 58.

35. S. Dillon Ripley, introduction to *Bundle of Feathers,* v–vi. Also see Futehally, "Portrait," 5–6.

36. Futehally, "Portrait," 5.

37. Ali, *Fall of a Sparrow,* 171.

38. Futehally, "Portrait," 5.

39. Sálim Ali, "The Nesting of the Baya (*Ploceus philippinus*). A New Interpretation of Their Domestic Relations," *BNHS* 34 (1931): 947–64.

40. Futehally, "Portrait," 5.

41. T. Anthony Davis, "Selection of Nesting Trees and Frequency of Nest Visits by Baya Weaverbird," in Ripley, *Bundle of Feathers,* 11–21.

42. Sálim Ali, "Flower-Birds and Bird-Flowers in India," *BNHS* 34 (1932): 573–605.

43. Gadgil, "Naturalist Extraordinaire," iv.

44. Interview with J. C. Daniel, February 12, 1998.

45. Ali, *Fall of a Sparrow,* 205–6.

46. See, for example, Kumar, *Science and the Raj;* Mary Louise Pratt, *Imperial Eyes: Travel Writing and Transculturation* (New York: Routledge, 1992).

47. Historically in India many aristocrats and royalty had observed and written about nature. Prominent in this tradition were the emperor Ashoka (third c. B.C.), who codified hunting restrictions for certain rare animals into law, and the sixteenth-century Mughal emperor Jehangir, who was a committed naturalist. Ali is significant, however, because of his writing in English (and thus with an orientation to an international audience), his engagement with scientific debates in the West, and his publication of guidebooks for the amateur Indian bird watcher—as opposed to royal writings meant only for other courtiers.

48. Prater did participate in the Constituent Assembly that drew up the constitution of independent India in 1946, so he was certainly not against Indian independence.

49. Sálim Ali, "Ornithology in India: Its Past, Present, and Future," *Proceedings of the Indian Science Academy* 33B, no. 3 (1971).

50. Ali writes that British ornithologists began to focus on Africa rather than India after 1947: "this is evident from the spate of papers on African ornithology appearing in the *Ibis* since that period, almost completely displacing India which formerly figured so largely in that journal." Ibid., 29.

51. This shift had begun earlier (see note 29 of this chapter). See also Golley, *Ecosystem Concept,* 2. As early as 1929 an American businessman, Arthur Vernay, had funded an ornithological expedition in India conducted by the BNHS. Ali, "Ornithology in India," 24.

52. As quoted in Ali, *Fall of a Sparrow,* 96, 248–49. After a rocky start, the two men eventually became friends (well after India's independence), and shortly before his death Meinertzhagen allowed Ali to read his unpublished diaries—these quotations are from Ali's notes at that time.

53. Ali, *Fall of a Sparrow,* 245.

54. SIA, RU 9591, int. 15, 346–79 (draft). There have been no book-length biographies of Ripley, but there is an extensive oral history of Ripley at the SIA. This document has provided me with much of the basic biographical information presented here.

55. SIA, RU 9591, int. 14, p. 324 (draft), emphasis in original; S. Dillon Ripley, "A New Race of Nightjar from Ceylon," *Bulletin of the British Ornithologists' Club* 65 (June 20, 1945): 40–41; Ripley int. 14, 371–73 (draft).

56. Ali to Ripley, August 22, 1943, SIA, RU 7008, CD 5, box 8, f. 12.

57. Ali technically did not have the title *Dr.,* never having received a single college degree. He did hold several honorary doctorates by the end of his life, however, and many people referred to him as Dr. Ali as a sign of respect. Ripley, "View from the Castle," 10; SIA, RU 9591, int. 18, p. 464 (draft).

58. Ali, *Fall of a Sparrow,* 179.

59. Ripley, "View from the Castle," 10; *Bundle of Feathers.*

60. Ali, *Fall of a Sparrow,* 179.

61. SIA, RU 9591, int. 2, p. 35 (draft); int. 13, p. 320 (draft).

62. Michael Werner, "The Office of Strategic Services: America's First Intelligence Agency," Central Intelligence Agency, <http://www.odci.gov/cia/publications/oss/art03.htm>.

63. SIA, RU 9591, int. 13, p. 321 (draft).

64. Ibid., int. 5, pp. 105–6, 106–7 (draft); Stresemann to Ripley, December 31, 1947, SIA, RU 7008, acc. 89–149, box 12, f. "Stresemann, Erwin."

65. SIA, RU 9591, int. 7, p. 150 (draft).

66. Ibid., int. 5, p. 117 (draft).

67. Ibid., p. 119 (draft).

68. Hutchinson's role in New Haven was similar to that of Stresemann in Berlin. He was a justifiably renowned scholar but even more important as a mentor of great ecologists, many of them American, like Robert MacArthur, cocreator of the island biogeography theory (see chapter 6). Hutchinson's career and influence is discussed at great length in Wilson, *Naturalist,* and Kingsland, *Modeling Nature.*

69. Kingsland's history of population ecology ignores behavioral biology and thus much of the German contribution to ecology before World War II, although she does discuss the crucial contributions made by Mayr at Harvard. She also relates how U.S. ecologists in the 1920s trained scientists from throughout the world, most notably some ecologists from the Soviet Union who developed an outstanding ecological tradition in the USSR that was unfortunately crushed by Lysenko and his political allies in 1939. Kingsland, *Modeling Nature,* 146–60.

70. S. Dillon Ripley, "The Smithsonian's Role in U.S. Cultural and Environmental Development," *BioScience* 36, no. 3 (March 1986): 155.

71. Mayr to Ripley, November 18, 1940, SIA, RU 7008, CD 1, box 6, f. 4, "Ernst Mayr."

72. Mayr to Ripley, December 12, 1944, SIA, RU 7008, CD 4, box 9, f. 3, "M."

73. Ali, *Fall of a Sparrow,* 179.

74. Ali to Ripley, September 9, 1944; November 23, 1945; SIA, RU 7008, CD 5, box 8, f. 12, "Sálim Ali."

75. "The Profession Honors S. Dillon Ripley," *Museum News* 64 (October 1985): 60–63.

76. This group included two close relatives who went on to hold the positions of curator and honorary secretary, Zafar Futehally and Humayun Abdulali, and the most long lasting and respected of his lieutenants, J. C. Daniel, current honorary secretary and ex-curator of the society.

77. Mayr to Ripley, December 12, 1944, SIA, RU 7008, CD 5, box 9, f. 3, "M."

78. The fine mesh of mist nets is practically invisible; ornithologists set them up on poles on edges of clearings or in forests. The nets catch but do not kill any small bird that happens to fly into them. Anyone willing to endure the occasional pinch of a beak can untangle the birds from the nets, weigh them, band them, and release them. Generally, the nets are rolled into a cord stretched between the poles and are unfurled only at dusk and dawn and left for a few minutes. This procedure can be repeated for several days in the same area to insure that an accurate sample is gathered. Mist nets were quite an innovation, allowing the capture of birds that were otherwise quite difficult to see or find. Previously almost all birds were collected by shooting them, and a rare few were captured by snares. These nets are discussed in great detail in Elliott McClure, *Bird Banding* (Pacific Grove, Calif.: Boxwood Press, 1984).

79. Ali, *Fall of a Sparrow,* 180.

80. The new work Ali was proposing was different than his popular handbook, *The Book of Indian Birds,* which did not include all the Indian species nor nearly as much detailed scientific information. In the first half of the twentieth century two major technical works had been undertaken in India, both by British scientists. Neither included Ali, or any other Indian, as coauthor. They also were limited to standard taxonomic information, not ecology.

81. See, for instance, E. O. Wilson, *Biophilia* (Cambridge, Mass.: Harvard University Press, 1984).

82. SIA, RU 9591, int. 18, pp. 453–54 (draft), emphasis in original.

83. Ali to Ripley, January 5, 1956, SIA, RU 7008, acc. 92–063, box 1, f. "Ali," emphasis in original.

84. A series is a collection of birds, all of the same species, that can be used to demonstrate the range of characteristics observed in the species. Mayr to Ripley, December 12, 1944, SIA, RU 7008, CD 5, box 9, f. 3, "M."

85. Ali to Ripley, December 7, 1947; August 10, 1949; June 13, 1949; July 12, 1949; January 5, 1956; SIA, RU 7008, acc. 92–063, box 1, f. "Ali."

86. Ali to Ripley, July 4, 1955, SIA, RU 7008, acc. 92–063, box 1, f. "Ali," emphasis in original.

87. When the tenth and final volume was published in 1974, Prime Minister Indira Gandhi officially released it. She also released the tenth edition of Sálim Ali's popular handbook *The Book of Indian Birds*.

88. Mahesh Rangarajan, "Five Nature Writers: Corbett, Anderson, Ali, Sankhala, and Krishnan," in *Environment and Wildlife: Five Essays, Research in Progress Papers, History and Society*, 3d series, no. 29 (New Delhi: Nehru Memorial Museum and Library, March 1998), 17; Pratt, *Imperial Eyes*, 6.

89. I can't help but note that the languages they shared were not just scientific—how different would this story have been if the Dutch or Portuguese had succeeded in colonizing India? The use of English as a language of education within India has no doubt influenced the close ties that subsequently developed with the United States as a source of ecological journals and scientists.

Chapter 3

1. Ali also had the good fortune to have met a very wealthy Singaporean businessman exiled to India during World War II. Loke Wan Tho became a good friend of Ali and an avid bird watcher and regularly gave money to the BNHS until his death, in 1964, after which his family continued donations in his memory. Ali, *Fall of a Sparrow*, 121–31.

2. Although there were Indian zoologists and botanists who were very active in the BSI and ZSI, they were more properly taxonomists than ecologists. Similarly, there were early forest officers who were dedicated naturalists and conservationists, but they also did not have much, if any, exposure to the new ecology that Ali learned in Germany. If there were any other ecologists with similar training and ability during this period, they were working far outside the main corridors of power, both within India and the international ecological community.

3. Ali to Ripley, March 25, 1987, SIA, RU 7008, acc. 95–085, box 1, f. "Ali."

4. This was not true with regard to all aspects of Indian ecology, but still there is not a comparable ornithologist in India who has taken Ali's place.

5. This is a relatively recent designation that reflects the number of species of plants and animals in India. Conservation International, an American NGO, publishes one of the more respected megadiversity listings, in which they include the top twenty nations (by number of species) in each of several categories, including plants, mammals, amphibians, reptiles, and birds.

6. Surprisingly, not much more is known about the jackal today. S. H. Prater, *The Book of Indian Animals*, 2d ed. (Delhi: Oxford University Press, 1971 [1948]), 127.

7. Mahesh Rangarajan, *Troubled Legacy: A Brief History of Wildlife Preservation in*

India, occasional paper, Nehru Memorial Museum and Library, New Delhi, 1988, 17. Also, George Schaller, *The Deer and the Tiger: A Study of Wildlife in India* (Chicago: Chicago University Press, 1967), 4–7.

8. Schaller, *Deer and Tiger,* 5.

9. Rangarajan, *Troubled Legacy,* 18.

10. Editorial note, Ashok Kothari and B. F. Chhapgar, eds., *Sálim Ali's India* (Delhi: Oxford University Press, 1996), 74. The last three cheetahs reported were shot at night from a car, by the same man who reported their sighting. Gee, *Wildlife,* 49–50.

11. Ali, *Fall of a Sparrow,* 143.

12. Annual report, *JBNHS* 69, no. 3 (1972): 691.

13. This process also helped keep the BNHS a vibrant organization. Ripley writes, "Asia has had its share of such societies, often founded by resident Westerners. Wars and Revolutions [political independence] have decimated many of these educational efforts . . . the one notable exception is Bombay." Ripley, "View from the Castle," 10.

14. "The Opening of Hornbill House: 13-3-1965," *JBNHS* 62, no. 1 (1965): 185; "Minutes of the Annual General Meeting," *JBNHS* 68, no. 3 (1971): 882; Ripley, "View from the Castle," 10.

15. Stephen Berwick and V. B. Saharia, eds., *The Development of International Principles and Practices of Wildlife Research and Management: Asian and American Approaches* (Delhi: Oxford University Press, 1995).

16. Interviews with A. J. T. Johnsingh, March 16, 20, 1998.

17. For one of many examples of Crichton's support of Schaller, see the preface to his novel *Congo* (New York: Knopf, 1980). Schaller holds a B.A. in anthropology and a B.S. in zoology from the University of Alaska and a Ph.D. in zoology from the University of Wisconsin.

18. Schaller's first book on the mountain gorilla had just been released that year to great acclaim: *The Mountain Gorilla: Ecology and Behavior* (Chicago: University of Chicago Press, 1963). While he was in India, a second book, more popular than scientific, was released: *The Year of the Gorilla* (Chicago: University of Chicago Press, 1964).

19. Schaller's books set a high standard. His popular books, such as *The Year of the Gorilla,* reached a larger audience than *The Deer and the Tiger.* Among his scientific books, his research project immediately following his work in India was a similarly focused study on the African lion and its associated prey species: *The Serengeti Lion: A Study of Predator-Prey Relations* (Chicago: University of Chicago Press, 1972). This book won the National Book Award for the sciences.

20. Sukumar, *Elephant Days.*

21. Schaller, *Deer and Tiger,* 8.

22. Robert Enders, review of *The Deer and the Tiger* by George Schaller, *American Scientist* 55, no.3 (Fall 1967): 306A–8A.

23. J. C. Daniel and B. R. Grubh, "The Indian Wild Buffalo, *Bubalus bubalis,* in Peninsular India: A Preliminary Study," *JBNHS* 63, no. 1 (1966): 32–53.

24. It was also very successful outside India, with favorable reviews in most ecological and popular science journals in the United States.

25. E. P. Gee, review of *The Deer and the Tiger* by George Schaller, *JBNHS* 64, no. 3 (1967): 530. E. P. Gee was a British tea planter from the state of Assam who stayed in India after 1947 and increasingly devoted his life to advocating environmental conservation measures there. Gee is credited with almost single-handedly preserving the rhino population in Kazaringa National Park, in Assam. (See Ali, *Fall of a Sparrow*, 189–92.)

26. Gee, review, 532.

27. "E. P. Gee: Obituary," *JBNHS* 66, no. 2 (1969): 361–64.

28. Thomas Kuhn, *The Structure of Scientific Revolutions* (Chicago: University of Chicago Press, 1962).

29. And further, it appears that many of the Ph.D.s who have done work in India first came to India as graduate students.

30. Schaller, *Deer and Tiger*, 357.

31. Spillett's work is summarized in his thesis. James Juan Spillett. "The Ecology of the Lesser Bandicoot Rat in Culcutta" (Calcutta: Bombay Natural History Society and Johns Hopkins University Center for Medical Research and Training, 1968).

32. Editorial, *JBNHS* 63, no. 3 (1966): 491.

33. Sálim Ali, review of *Wildlife of India* by E.P. Gee, *JBNHS* 61, no. 2 (1964): 426.

34. Editorial, *JBNHS* 63, no. 3 (1966): 489.

35. J. J. Spillett, "General Wild Life Conservation Problems in India," *JBNHS* 63, no. 3 (1966): 617.

36. Ali to Ripley, May 5, 1949; May 23, 1949; SIA, RU 7008, acc. 92–063, box 1, f. "Ali."

37. *Wilderness*, or *wild areas*, is used here to mean a landscape ostensibly without people, or without visible human modifications, and is thought by U.S. ecologists to be a precondition for good field research. For more on wilderness, see Roderick Nash, *Wilderness and the American Mind*, 3d ed. (New Haven: Yale University Press, 1982).

38. Annual report, *JBNHS* 69, no. 3 (1972): 691.

39. Alexander Adams, ed., *First World Conference on National Parks* (Washington, D.C.: U.S. Department of the Interior, 1962), 413.

40. There were three thousand members in 1966. Ecology was still a small subfield of biology, representing less than one percent of the 1965 graduate student enrollments in biology in 1965. This data is all from Chunglin Kwa, "Representations of Nature Mediating between Ecology and Science Policy: The Case of the International Biological Programme," *Social Studies of Science* 17 (1987): 417.

41. As cited in Kingsland, *Modeling Nature*, 244n12. Eugene Odum, the author of the best-selling textbook *Fundamentals in Ecology* (Philadelphia: Saunders, 1953), the first explicitly ecology textbook, charts this growth in interest in ecology by the ballooning sales of his textbook, with its third edition published in 1971.

42. Wilson, *Naturalist,* 356.

43. Mary Louise Pratt discusses this Linnaean missionary zeal, linking it to European colonialism. Pratt, *Imperial Eyes,* ch. 1.

44. William Cronon, "The Trouble with Wilderness, or Getting Back to the Wrong Nature," in *Uncommon Ground,* ed. William Cronon (New York: Norton, 1995), 76.

45. For early examples of these trends, see Henry Nash Smith, *Virgin Land* (Cambridge, Mass.: Harvard University Press, 1950); Nash, *Wilderness.*

46. Cronon, "Trouble with Wilderness," 78.

47. See "Theodore Roosevelt: Manhood, Nation, and 'Civilization,'" in Gail Bederman, *Manliness and Civilization: A Cultural History of Gender and Race in the United States* (Chicago: University of Chicago Press, 1995), 170–215.

48. There is a vast literature dealing with these points, including many debates surrounding Cronon's arguments. A useful primer for these debates is J. Baird Callicott and Michael Nelson, eds. *The Great New Wilderness Debate* (Athens: University of Georgia Press, 1998).

49. For more on Clement's theory and the difficulty with which it was displaced in the 1950s, in spite of voluminous conflicting evidence, see Michael Barbour, "Ecological Fragmentation in the Fifties," in *Uncommon Ground,* ed. William Cronon (New York: Norton, 1995), 233–55.

50. Cronon, "Trouble with Wilderness," 82.

51. Again, see Mary Pratt for more on this construction of ecological frontiers.

52. Interview with Ullas Karanth, Bangalore, July 7, 1998.

53. Conference on the Future of the Smithsonian Institution, "Some Achievements during Eighty Years," in *Proceedings of the Conference on the Future of the Smithsonian Institution, February 11, 1927* (Baltimore: Lord Baltimore Press, 1927), 6; C. G. Abbot, "The Smithsonian Institution—Its Activities and Capacities," in *Proceedings,* 24. For the frantic communications of the Smithsonian with their field representatives in Panama, see SIA, RU 134, box 1 f. 5.

54. Paul Oehser, *The Smithsonian Institution,* 2d rev. ed. (Boulder: Westview Press, 1983), 99–100.

55. In 1982 federal funding alone contributed $2.7 million to the operation of BCI, not including private grants researchers brought with them. Oehser, *Smithsonian,* 100.

56. Ibid.

57. My understanding of BCI was supplemented by numerous interviews with scientists who worked there, especially Renee Borges, John Nason, and Vidya Athreya. The history of BCI is also discussed in Wilson, *Naturalist.* In the last ten years, BCI has begun to change its previous isolationist policies. See Catherine Christen, "At Home in the Field: Smithsonian Tropical Science Field Stations in the U.S. Panama Canal Zone and the Republic of Panama," *Americas* 58, no.4 (April 2002): 537–75.

58. Wilson, *Naturalist,* 196. The idea of "becoming a man" is also a common theme within histories of colonialism, in which young men working for the imperial state are often understood by historians within the context of gender ideologies. For more on

this theme, see, for example, Anne McClintock, *Imperial Leather: Race, Gender, and Sexuality in the Colonial Conquest* (New York: Routledge, 1995). Also instructive is Van Gosse, who argues that some young U.S. men in the fifties turned to a specific type of radical, adventurous, and exotic politics—those of Fidel Castro in the jungles of Cuba—as a masculinity-affirming activity. Gosse, *Where the Boys Are: Cuba, Cold War America, and the Making of the New Left* (New York: Verso, 1993).

59. For accounts of women naturalists in the United States, see Marcia Myers Bonta, *Women in the Field: America's Pioneering Women Naturalists* (College Station: Texas A&M University Press, 1991). I suspect that women worked near home because of the social limitations on women traveling alone "in the wild," as opposed to the many women who successfully carved out a sphere for independent action in the cities or at least moderately populated areas, as with Hull House, anthropological work like that of Margaret Mead, or missionary activities in the colonies. All these activities would offer opportunities for adventure, travel, and autonomy.

60. Ali, *Fall of a Sparrow,* 182.

61. Striking evidence of the difficulties women had in finding academic employment in lab-based science departments is offered by Margaret Rossiter, *Women Scientists* (Baltimore: Johns Hopkins University Press, 1982). I believe that this gender bias was, and is, even more pronounced in tropical field ecology. See Pam Henson, "Invading Arcadia: Women Scientists in the Field in Latin America, 1900–1950," *Americas* 58, no. 4 (2002): 577–600.

62. Women primatologists are discussed in more detail in Donna Haraway, *Primate Visions* (New York: Routledge, 1989).

63. Wilson, *Naturalist,* 210.

64. For similar feelings about solitude, see George Schaller, *Golden Shadows, Flying Hooves* (New York: Random House, 1973).

65. For further examples of ecologists as explorers see, Jonathan Maslow, *Footsteps in the Jungle: Adventures in the Scientific Exploration of the American Tropics* (Chicago: Ivan R. Dee, 1996); Elliott McClure, *Stories I Like to Tell: An Autobiography* (Camarillo, Calif.: Elliott McClure, 1995).

66. See E. O. Wilson, *Consilience* (New York: Vintage Books, 1998).

Chapter 4

1. The official name of the wetlands is Keoladeo Ghana National Park, but it is most often referred to by the name of the nearby town, Bharatpur, a usage I will follow. (The wetlands are right on the edge of the town, and you can walk from town to the interior of the national park in a matter of a few minutes.)

2. This figure is given by Kailash Sankhala, "Livestock Grazing in India's National Parks," in *People and Protected Areas in the Hindu–Kush–Himalaya: Proceedings of the International Workshop on the Management of National Parks and Protected Areas in the Hindu–Kush –Himalaya, 6–11 May, 1985,* ed. J. A. McNeely, J. W. Thorsell, and S. R. Chalise (Kath-

mandu: King Mahendra Trust for Nature Conservation and International Centre for Integrated Mountain Development, 1985), 57.

3. Henry Kissinger, *White House Years* (Boston: Little, Brown, 1979), 845.

4. Judith M. Brown, *Nehru: Profiles in Power* (Harlow, England: Pearson Education, 1999), 128.

5. H. G. Alexander to Ripley, December 12, 1952, SIA, RU 7008, acc. 87–075, box 1, f. 1.

6. Ali to Ripley, November 23, 1945, SIA, RU 7008, CD 5, box 8, f. 12, "Sálim Ali." He could have just as easily been writing about India—their correspondence was often centered around attempts to get permission from the government of India to travel to out-of-the-way places.

7. Geoffrey T. Hellman, "Curator Getting Around," *New Yorker,* August 26, 1950, 31–49.

8. The article went on to give examples of Ripley's ability to "get on in that part of the world" that are disturbingly colonial in their tenor, as when Ripley complained about his accommodations in Nepal (which were actually adequate) so that his Nepalese hosts would think he was a very important person—to fail to complain would give the message that he did not deserve any better, he claimed.

9. Ali to Ripley, October 31, 1950, SIA, RU 7008, acc. 92–063, box 1, f. "Ali."

10. Ali to Ripley, December 10, 1950.

11. Ali to Ripley, January 18, 1951.

12. SIA, RU 9599, David Challinor interviews, int. 2, p. 64 (draft).

13. Interview with David Challinor, Washington, D.C., March 16, 2001.

14. McClure, *Bird Banding,* 2.

15. Ehrlich, Dobkin, and Wheye, *Birder's Handbook,* 95.

16. Ali, *Fall of a Sparrow,* 143; McClure, *Bird Banding,* 2.

17. Ali, *Fall of a Sparrow,* 143.

18. Ibid., 144.

19. This means that if scientists at the CDC were to work with a live strain of this virus, they would do so in a completely restricted and air-tight environment, including "space" suits and sealed rooms.

20. Telford H. Work and H. Trapido, "Summary of Preliminary Report on Investigation of the Virus Research Centre on an Epidemic Disease Affecting Forest Villagers and Wild Monkeys of Shimoga District," *Indian Journal of Medical Science* 11 (1957): 340–41.

21. Ali, *Fall of a Sparrow,* 144. See also K. Banerjee, "Emerging Viral Infections with Special Reference to India," *Indian Journal of Medical Research* 103 (April 1996): 177–200.

22. One sample newspaper article: Mike Cooper, "U.S. Confirms West Nile Virus Caused N.Y. Deaths," Reuters Press, October 21, 1999.

23. Ali to Ripley, February 25, 1959, SIA, RU 7008, acc. 92–063, box 1, f. "Ali."

24. The Virus Research Centre in Pune has led studies of the KRD virus since it was identified by Telford Work. The VRC continues to conduct research on this virus See Banerjee, "Viral Infections."

25. Ripley to Ali, March 12, 1959, SIA, RU 7008, acc. 92–063, box 1, f. "Ali."

26. The history of the WHO program is described in Ali to Indira Gandhi, October 6, 1972, SIA, RU 7008, acc. 92–063, box 1, f. "Ali."

27. Ali, *Fall of a Sparrow,* 144.

28. Many ecologists who read earlier drafts of my manuscript objected to the implication that funding is easily obtained today. They feel there is a continued bias against field ecology on all but the most charismatic of species (such as tigers or lions), as compared to more industry-friendly sciences such as medical biology, chemistry, or physics. Some ecologists in the United States are actively attempting to create a National Institute of the Environment, like the National Institute of Health, which could fund ecological studies in the United States and address this imbalance. Still, I maintain that compared to the fifties and sixties, ecologists today live in a golden age of funding and access to resources, with dozens of private and governmental funding programs available for grants. And after all, does any researcher (scientist or historian) ever feel there is enough money?

29. Of course, this program was ultimately unsuccessful in eradicating the disease, though many areas showed dramatically decreased malaria rates. Money figures come from Javed Siddiqi, *World Health and World Politics* (Columbia: University of South Carolina Press, 1995), 225.

30. Ali, *Fall of a Sparrow,* 145.

31. Since the United States and Canada started keeping joint bird-banding records in 1920, over 40 million birds have been banded. Of those only five percent were ever seen again. For even more information on the techniques and uses of this method, see McClure, *Bird Banding.* Ehrlich, Dobkins, and Wheye *The Birder's Handbook,* also provides a useful history of bird-banding, including the figures on U.S. recapture rates.

32. Its reemergence in the eighties was attributed to a tick population explosion in Karnataka linked to the clearing of forests. Reported in Banerjee, "Viral Infections." For a lengthier discussion of this reemergence, see World Resources Institution, "Land Conversion," <http://www.wri.org/wri/wr-98–99/landconv.htm>.

33. Ali, *Fall of a Sparrow,* 145; also, Ali to Indira Gandhi, October 6, 1972, SIA, RU 7008, acc. 92–063, box 1, f. "Ali."

34. Ali, *Fall of a Sparrow,* 145.

35. Ali to Indira Gandhi, October 6, 1972.

36. Obviously, these are (mainly) the nations of SEATO—after 1966 the MAPS offices were physically moved to Bangkok, where they were attached to SEATO's Applied Scientific Research Corporation.

37. The metal band bore a Hong Kong address for return to MAPS. McClure, *Stories.*

38. Ibid. Also see Audubon Society, "Obituary: Dr. H. Elliott McClure, April 29, 1910-December 27, 1998," <http://conejovalley.ca.audubon.org/elliot.html>.

39. Ali, *Fall of a Sparrow,* 145.

40. Bryon W. Brown, "The Use of U.S.-Owned Excess Foreign Currencies," un-

published report to the U.S. Advisory Commission on International Educational and Cultural Affairs, SIA, RU 274, box 11, f. 4, p. 23. I rely heavily on Brown's summary for my information throughout this discussion of the PL 480 funds.

41. SIA, RU 9591, David Challinor Interviews. int. 2, p. 45 (draft). Also "The Smithsonian Institution Foreign Currency Program in Archaeology and related Disciplines," SIA, RU 271, box 8, f. 1.

42. Ripley to Leland J. Haworth, August 24, 1965, SIA, RU 271, box 8, f. 1.

43. Kennedy B. Schmertz to Ali, June 23, 1966, SIA, RU 7008, acc. 92–063, box 1, f. "Ali."

44. George E. Watson to Kennedy Schmertz, March 15, 1968, memorandum, SIA, RU 271, box 8.

45. The transcript of this NBC television show was entered into the *Congressional Record,* March 24, 1969, S3141–44.

46. Herman Schaden, "Bird Project a Coverup for War Test, NBC Says," *Washington (D.C.) Evening Star,* February 4, 1969, A-1.

47. Two POBS scientists discuss this lack of coordination: "It is unfortunate that Dr. McClure who has been advised in a general way of our large banding operations in the Central Pacific did not let us know that his organization was planning a widespread publicity program and it is equally unfortunate that we did not advise him of ours." Philip Humphrey to Tom Harrison, March 17, 1965, SIA, RU 245, box 19, f. 17.

48. Philip M. Boffey, "Biological Warfare: Is the Smithsonian Really a Cover?" *Science* 163 (1969): 791–96.

49. For an account of the Army's goals with the POBS study, see Ed Regis, *The Biology of Doom: The History of America's Secret Germ Warfare Project* (New York: Holt, 1999), 188–92. Another account is found in Ted Gup, "The Smithsonian's Secret Contract: The Link between Birds and Biological Warfare," *Washington Post Magazine,* May 12, 1985, 10–17.

50. Graham S. Pearson, "BTWC Security Implications of Human, Animal, and Plant Epidemiology." *Strengthening the Biological Weapons Convention,* Briefing Paper no. 23, Report of the NATO Advanced Research Workshop, Cantacuzino Institute, Bucharest, June 3–5, 1999 (Brussels: NATO, 1999).

51. McClure, *Stories,* 280–297.

52. Galler quoted in Gup, "Smithsonian's Secret Contract," 17.

53. James Jones, *Bad Blood* (New York: Free Press, 1981).

54. Drew Featherstone and John Cummings, "CIA Linked to 1971 Swine Virus in Cuba," *Washington Post,* January 9, 1977, A-2.

55. A sensationalist but thorough account of these tests is found in Robert Harris and Jeremy Paxman, *A Higher Form of Killing: The Secret Story of Chemical and Biological Warfare* (New York: Hill and Wang/Noonday Press, 1982).

56. Boffey, "Biological Warfare," 791.

57. K. C. Emerson to William E. Small, November 1, 1968, SIA, POBS Research File, The Pacific Ocean Biological Survey Program.

58. Ripley to Ali, September 25, 1969, SIA, RU 7008, acc. 92–063, box 1, f. "Ali."

59. Gup, "Smithsonian's Secret Contract," 12.

60. Smithsonian Institution, *Pacific Ocean Biological Survey Program Annual Report for the Year Ended June 30, 1969,* 1 September 1969, stored in SIA, RU 245, box 15, f. 3, pp. 1–27.

61. Nixon's actions were spurred not just by the POBS allegations but also by the famous sheep incident of March 1968, in which thousands of sheep were killed by nerve gas that escaped from a field test at the Dugway Proving Ground in Utah. Philip M. Boffey, "Nerve Gas: Dugway Accident Linked to Utah Sheep Kill," *Science,* 162 (1968): 1460–64. This incident is also covered, though less ably, in Regis, *Biology of Doom,* 207–10.

62. Kennedy Schmertz to Ripley, August 3, 1970, memorandum, SIA, RU 271, box 8.

63. George E. Watson to David Challinor, February 13, 1969, memorandum, SIA, RU 7008, acc. 92–063, box 1, f. "Ali"; *Congressional Record,* March 24, 1969, S3143; Ripley to Ali, February 28, 1969; September 25, 1969; SIA, RU 7008, acc. 92–063, box 1, f. "Ali."

64. Ripley to Ali, September 10, September 25, 1969, SIA, RU 7008, acc. 92–063, box 1, f. "Ali."

65. M. V. Kamath, "Sponsoring Research in India: Pentagon Assailed," *Times of India,* August 16, 1969; Ali to Ripley, October 7, 1969, SIA, RU 7008, acc. 92–063, box 1, f. "Ali."

66. Kissinger, *White House Years,* 916. For a detailed firsthand account of the U.S. response to the 1971 India-Pakistan War, see pp. 842–918.

67. "Rupee Agreement Advantageous to India and USA, Says Mr. Moynihan," *India News* (Information Service, Embassy of India), February 15, 1974, p. 1; Ali to Indira Gandhi, October 6, 1972, SIA, RU 7008, acc. 92–063, box 1, f. "Ali."

68. Ali, *Fall of a Sparrow,* 146; Ali to Ripley, May 17, 1974, SIA, RU 7008, acc. 92–063, box 1, f. "Ali"; Sálim Ali, "Baseless Charges," letter to the editor, *Times of India,* August 14, 1974.

69. This study involved infecting terminal cancer patients with KFD virus, and another virus, to work on developing vaccines for these and other tick-transmitted diseases. The involvement of Porton Down scientists made it appear to some observers that the study had military implications, whether in the development of a vaccine or for attack. The published study is strictly scientific: Webb et al., "Leukemia and Neoplastic Process Treated with Langat and Kyasanur Forest Disease Viruses: A Clinical and Laboratory Study of Twenty-Eight Patients," *British Medical Journal,* January 29, 1966, 258–66.

70. This warning list is available at: <http://www.acda.government./factshee/wmd/bw/auslist.htm>.

71. The years of the MAPS program precisely overlay the Vietnam War: 1963–75.

72. For the complete story see Regis, *Biology of Doom.*

73. McClure, *Stories,* 283. McClure reports that scientists at the Russian Bureau of Bird Banding were very helpful and consistently forwarded their information to him and his project.

74. Ali, *Fall of a Sparrow,* 146–7.

75. The Indian parliament, the Lok Sabha, published a report on the Indian Migratory Bird Project in April 1975 by the Public Accounts Committee, which summarized its findings. I have used secondary accounts found in the Smithsonian archives and Ali's autobiography.

76. Ali, *Fall of a Sparrow,* 145.

77. The story was told to me by several ecologists there and is also related in Chris Wemmer and K. Ullas Karanth, "Reflections on Fieldwork Abroad: Attitudes and Latitudes: Observations and Platitudes," *Grapevine* 39 (March 2001): 3–4. The quote at the beginning of this section was used by Ripley in complaint about constant Indian suspicions of his bird projects. Ripley to Daniel Moynihan, August 29, 1974, SIA, RU 254, box 100, f. 3.

78. For more on this story and its links to the politics of Project Tiger, see chapter 7.

79. Madhav Gadgil and Ramachandra Guha, *Ecology and Equity: The Use and Abuse of Nature in Contemporary India* (New Delhi: Penguin, 1995), 112.

Chapter 5

1. Interview with Madhav Gadgil, Indian Institute of Science, Bangalore, July 9, 1998. Much of the material in the pages about Gadgil comes from this interview.

2. Madhav Gadgil, "In Love with Life," *Seminar 409: Our Scientists,* September 1993, 27. Page numbers in the following section are from this work.

3. This is fairly unusual—Gadgil was the first person in this particular lecture series to do so. Most academics lecture either in English or, more rarely, in Hindi. Mahesh Rangarajan, pers. comm., March 15, 2001.

4. See Kingsland, *Modeling Nature.*

5. Wilson describes this course in his autobiography as "focused on basic theory. At first I thought I had failed by pushing model construction too far at the expense of natural history. . . . I came to realize later that many of the students were greatly influenced by my presentation, and a few were drawn into population biology as a career." Wilson, *Naturalist,* 256–57.

6. It is possible to overstate their affluence—both men came from wealthy families, Ali more so than Gadgil, but by no means was either man positioned in the Indian industrial-royal elite.

7. These included William Bossert, Ross Kiester, Daniel Simberloff, Robert Trivors, Jonathan Roughgarden, Thomas Schoener, and Joel Cohen, among many others. Princeton, Yale, and Chicago also were vibrant centers of ecological study.

8. This idea was, and is, quite controversial and is the basis of both Wilson's

"sociobiology" and Gadgil's atrocious analysis of the Indian caste system as a naturally evolved partition of resources in *This Fissured Land*.

9. Gadgil continues to have an impact on U.S. ecologists long after he left to return to India. One example: in the introduction to a widely read monograph (it was the subject of a semester-long graduate seminar in ecology at the University of Iowa in the spring of 1999), the author wrote that he "extends his appreciation to James Karr and Madhav Gadgil for posing some of the stimulating questions and ideas back in the early 1970s that initiated this research." Warren G. Abrahamson and Arthur E. Weis, *Evolutionary Ecology across Three Trophic Levels: Goldenrods, Gallmakers, and Natural Enemies* (Princeton: Princeton University Press, 1997), xi.

10. Hugh Finley et al., eds., *Lonely Planet Travel Survival Kit: India* (Oakland: Lonely Planet Publications, 1996), 885.

11. CES, *A Decade of Ecological Research and Training, 1983–1993* (Bangalore: Centre for Ecological Sciences, IISc, 1993), 1. The CES was supported by the newly formed Department of the Environment (DoE) of the government of India, a precursor to the Ministry of Environment and Forests, as a "Centre of Excellence" with a mandate to study the Western Ghats, a site of high endemicity and one of the global hot spots of biodiversity. The five-year lag between the proposal in 1978 and the formation in 1983 was partially attributable to the formation of the Department of the Environment—the original proposal had been sent to the Department of Science and Technology (DST); the DST decided that the soon-to-be-formed DoE would be better suited for the proposed CES, and thus had delayed the approval of the proposal until the DoE was up and running.

12. Wilson made his initial scientific reputation for his work on social insects, particularly ants.

13. Gadagkar interview, Indian Institute of Science, Bangalore, July 1, 1998.

14. Joshi interview, Indian Institute of Science, Bangalore, July 9, 1998.

15. This was especially evident in interviews with Ravi Chellam of the Wildlife Institute of India and many of the scientists at the BNHS, including the director, Asad Rahmani—strikingly, the scientists who most freely made these statements were typically outstanding in their own right, but they perceive a difference in the type of ecology practiced at CES.

16. The so-called Science Wars of the nineties have so polarized scientists and science studies scholars that often even relatively benign statements are rejected from fear of the "give them an inch, they'll take a mile" variety. See Andrew Ross, ed., *Science Wars* (Durham: Duke University Press, 1996).

17. See, for example, Ashish Nandy, ed., *Science, Hegemony, and Violence: A Requiem for Modernity* (Delhi: Oxford University Press, 1988); Vandana Shiva, *Monocultures of the Mind: Perspectives on Biodiversity and Biotechnology* (Dehradun: Natraj Publishers, 1993); Lesley Instone, *Science, Technology, and Western Domination: Some Aspects of Cultural Imperialism in the Third World* (Melbourne: Monash University, 1985).

18. Such allegations have not been made in print. Even Ram Guha, who has been

quite critical of the effect of U.S. ecology as an imperial force in India, holds up the scientists of the CES as models for a better future in "The Authoritarian Biologist and the Arrogance of Anti-Humanism: Wildlife Conservation in the Third World," *Ecologist* 27, no. 1 (January–February 1997): 14–20. Guha has been a fellow at CES and has coauthored two books with Madhav Gadgil. Still, the CES is certainly open to such claims given its institutional history and the personal connections of its scientists. In personal conversations with three informants in India, this claim was made.

19. See, for example, David Arnold, *Colonizing the Body: State Medicine and Epidemic Disease in Nineteenth-Century India* (Berkeley: University of California Press, 1993), 116–58.

20. I chose to focus on Sukumar and Gadagkar among the four other faculty at CES, as Joshi admits that he is first and foremost a mathematical collaborator with other scientists, and Renee Borges had just joined the CES when I was there in 1998.

21. Gadagkar's recent publications include "The Origin and Evolution of Caste Polymorphism in Social Insects," *Journal of Genetics* 76 (1997): 167–79; "Do Social Wasps Choose Nesting Strategies Based on Their Brood Rearing Abilities?" *Naturwissenschaften* 84 (1997): 79–82; "Colony Founding in the Primitively Eusocial Wasp, *Ropalidia marginata* (Lep.) (Hymenoptera: Vespidae)," *Ecological Entomology* 20 (1995): 273–82; "Regulation of Worker Activity in a Primitively Eusocial Wasp, *Ropalidia marginata,*" *Behavioral Ecology* 6 (1995): 117–23; "Why the Definition of Eusociality Is Not Helpful to Understand Its Evolution and What We Should Do about It," *Oikos* 70 (1994): 485–88. (Eusocial is a scientific term; the wasps are commonly called "social" insects.)

22. Raghavendra Gadagkar, *Survival Strategies* (Cambridge, Mass.: Harvard University Press, 1997). Wilson's remarks appear on back of the dust jacket.

23. Among his many articles, one of the most accessible summaries is Gadagkar, "The Pains and Pleasures of Doing Ethology in India," in *Readings in Behaviour,* ed. R. Ramamurthi and Geethabali (New Delhi: New Age International, 1996), 8–9.

24. Raghavendra Gadagkar, "Western Scientists Set the Trends," *Down to Earth* 2, no. 7 (August 31, 1993): 45.

25. Ibid. Remember that Gadagkar gave up molecular biology himself. A dramatically different reading of Indian scientific work is possible. Kapil Raj argues that Indian scientists value a different sort of work, work more oriented to definitively working out all the details of a problem than to pushing research agendas in new directions. Raj, "Images of Knowledge, Social Organization, and Attitudes to Research in an Indian Physics Department," *Science in Context* 2, no. 2 (1988): 317–49. Raj did his work at IISc, with physicists. His article, at least, suggests that the CES has internalized American research values and that when Gadagkar argues for more creative science, he is arguing for more American science. Gadagkar is not alone, however, in arguing that Indian science is often derivative, second-rate, and dependent on Western science for its models. See Arnab Rai Choudhuri, "Practicing Western Science Outside the West: Personal Observations on the Indian Scene," *Social Studies of Science* 15 (1985): 475–505.

26. Terry Erwin, "Tropical Forest Canopies: The Last Biotic Frontier," *Bulletin of the Entomological Society of America* 29 (1983): 14–19. Other species were also affected, but most vertebrates could escape.

27. CES, *Ecological Sciences Research and Training Centre at the Indian Institute of Science Bangalore: Annual Report for the Period 1-4-90 to 31-3-91* (Bangalore: Centre for Ecological Sciences, IISc, 1991), 44. The percentage estimate comes from my interview with Gadagkar, currently the CES chairman.

28. Raghavendra Gadagkar, "Biology in the Twenty-First Century: Back to Stamp Collection?" *Scampus,* Spring 1998, 5. Many ecologists in India and the United States refer to taxonomy studies as stamp collecting, or butterfly collecting, implying that the work is not science but merely a hobby. The history of this insult is given in Kingsland, *Modeling Nature.*

29. E. O. Wilson, *Biophilia* (Cambridge, Mass.: Harvard University Press, 1984), 137–38.

30. One of the most extensive of these criticisms is Visvanathan, *Carnival for Science.*

31. Gadagkar, "Biology in the Twenty-First Century," 5.

32. Interview with R. Sukumar, IISC, Bangalore, June 12, 15, 1998.

33. Sukumar, *Elephant Days,* 119.

34. The definitive version of his research was published in book form in the Cambridge Studies in Applied Ecology and Resource Management series: *The Asian Elephant: Ecology and Management* (New York: Cambridge University Press, 1989). His popular-press book, *Elephant Days and Nights,* provides a more general overview. See also "Elephant-Man Conflict in Karnataka," in *Karnataka–State of Environment Report, 1984–85,* ed. C. J. Saldanha (Bangalore: Centre for Taxonomic Studies, 1986). For a briefer scientific summary of his arguments, see "The Management of Large Mammals in Relation to Male Strategies and Conflicts with People," *Biological Conservation* 55 (1991): 93–102.

35. There had been a previous study of crop raiding in Kenya, but it had been limited to an economic recounting of what had been lost.

36. In counterpoint to this, domesticated male elephants are put on starvation rations when they enter musth (and are uncontrollable and thus useless as a working animal) which hastens their return to hormonal normalcy.

37. Reviews are found in *Animal Behaviour* 48, no. 2 (August 1994): 492–93; *Journal of Wildlife Management* 58, no. 2 (April 1994): 383–85; *Quarterly Review of Biology* 66, no. 1 (March 1991): 93–94.

38. These include *Animal Behavior, Science, Biological Conservation, Nature, Journal of Ecology, Oecologia, Journal of Biogeography, Proceedings of the National Academy of Science* (U.S.), *American Naturalist,* and the popular U.S. magazine, *Natural History.*

39. Sukumar, "Management of Large Mammals."

40. Sukumar, *Elephant Days.* This theme is throughout the book, but especially the introduction and conclusion.

41. Fuller's work includes *Philosophy, Rhetoric, and the End of Knowledge: The Coming of Science and Technology Studies* (Madison: University of Wisconsin Press, 1993) and *Philosophy*

of Science and Its Discontents (Boulder: Westview Press, 1989). The Iowa Project on Rhetoric of Inquiry is an interdisciplinary group oriented around common questions of how arguments are constructed and made in the different branches within the academy: what constitutes effective writing or pedagogy, what counts as evidence, or as persuasive logic.

42. Anecdotal information on Hubbell's research is derived from interviews with various scientists who have worked on this project and from the published record. Hubbell and his students and colleagues have published widely and voluminously on this project. A sample is Richard Condit, Stephen Hubbell, and Robin Foster, "Mortality Rates of Two Hundred Five Neotropical Trees and Shrub Species and the Impact of a Severe Drought," *Ecological Monographs* 65, no. 4 (November 1995): 419–35.

43. Sukumar, *Elephant Days,* 140.

44. Raman Sukumar, "Dynamics of a Tropical Deciduous Forest: Population Changes (1988 through 1993) in a Fifty-Hectare Plot at Mudumalai, Southern India," in *Forest Biodiversity Research, Monitoring, and Modeling: Conceptual Background and Old World Case Studies,* ed. F. Dallmeier and J. A. Comiskey (Delhi: Parthenon Publishing, 1997), 529–40.

45. Interview with a U.S. scientist who has chosen to remain anonymous, May 6, 1999.

Chapter 6

1. In the state of Maharashtra alone, over three thousand sacred groves, from a small clump of trees to several hectares, have been identified thus far. For a more thorough introduction to sacred groves, see Madhav Gadgil and V. D. Vartak, "Sacred Groves of India: A Plea for Continued Conservation," *JBNHS* 72, no. 2 (1974): 313–20.

2. This argument is made in many places. See, for example, Gadgil and Guha, *Ecology and Equity;* Ghazala Shahabuddin, Ranjit Lab, and Pratibha Pande, *Small and Beautiful: Sultanpur National Park* (New Delhi: Kalpavriksh, 1995); Shankar Barua, "Of Pocket Sanctuaries: A Concept, a Proposal to Monitor a Small Lake near Delhi" (manuscript in WWF-India Library, n.d.).

3. Shekar Singh, *Biodiversity Conservation through Ecodevelopment: Planning and Implementation Lessons from India.* South-South Cooperation Programme on Environmentally Sound Socio-Economic Development in the Humid Tropics, Working Papers, no. 21 (Paris: UNESCO, 1997).

4. In 1991, India's population of approximately 950 million included 52 million tribal people.

5. Slightly fewer than twenty-five thousand people have been moved from all sanctuaries between 1968 and 1995—this figure does not include those people who encroach on the sanctuaries. Mahesh Rangarajan, pers. comm. Also see Ashish Kothari, P. Pande, S. Singh, and D. Variava. *Management of National Parks and Sanctuaries in India: A Status Report* (New Delhi: Environmental Studies Division, Indian Institute of Public Administration with the Ministry of Environment and Forests, 1989).

6. David Quammen's *The Song of the Dodo: Island Biogeography in an Age of Extinction* (New York: Simon and Schuster, 1996) is an award-winning book about the science and scientists of island biogeography and the extinction crisis of the late twentieth century. See also Grove, *Green Imperialism*.

7. See, for example, P. J. Darlington, *Zoogeography: The Geographical Distribution of Animals* (New York: Wiley, 1957); Ernst Mayr, "The Zoogeographic Position of the Hawaiian Islands," *Condor* 45 (1943): 45–48; F. W. Preston, "The Canonical Distribution of Commonness and Rarity, parts 1 and 2," *Ecology* 43 (1962): 185–215, 410–32.

8. Robert MacArthur and E. O. Wilson, "An Equilibrium Theory of Insular Zoogeography," *Evolution* 17 (December 1963): 373–87. The next article in *Evolution* cited thirty-eight other scientists, a much more typical number within that publication. Interestingly, these eleven scientists included Erwin Stresemann of Germany (Sálim Ali's mentor) and Ernst Mayr, suggesting that island biogeography is an extension of the new ecological work of the German behavioral ecologists of the early twentieth century.

9. R. H. MacArthur and E. O. Wilson, *The Theory of Island Biogeography* (Princeton: Princeton University Press, 1967). Note the change in their key phrase from "Insular Zoogeography" to "Island Biogeography."

10. In my review of this literature, I have not found any articles that failed to cite MacArthur and Wilson. For precise citation figures, see Quammen, *Song of the Dodo,* 434–36.

11. Edward Grumbine, *Ghost Bears: Exploring the Biodiversity Crisis* (Washington, D.C.: Island Press, 1992), 29. New theories, like metapopulations, have continued to refine island biogeography in the 1990s.

12. Wilson, *Naturalist,* 355.

13. The mathematical element of this theory was crucial, coming at a time when ecologists were attempting to increase the rigor of ecology (as also seen in molecular biology) and distance it from "natural history." This is illustrated anecdotally in Wilson's autobiography and is more historically conceptualized in Kingsland, *Modeling Nature.*

14. Stephen Budiansky offers a scathing critique of island biogeography, and especially the Species-Area Curve, in *Nature's Keepers: The New Science of Nature Management* (New York: Free Press, 1995).

15. Grumbine, *Ghost Bears,* 29.

16. Malcolm Hunter, *Wildlife, Forests, and Forestry: Principles of Managing Forests for Biological Diversity* (Englewood Cliffs, N.J.: Prentice-Hall, 1990), 115.

17. Jared Diamond, "The Island Dilemma: Lessons of Modern Biogeographic Studies for the Design of Natural Reserves," *Biological Conservation* 7 (1975): 129–46. Diamond refers to MacArthur and Wilson's work as "a recent scientific revolution" (131) and throughout has the tone of a devoted apostle expanding and carrying the word to the masses, Paul to MacArthur and Wilson's Jesus.

18. Edge effect had been noted decades earlier, particularly in logged forests. Two early studies are Rudolf Geiger, *The Climate near the Ground,* trans. Scripta

Technica, Inc. (Cambridge, Mass.: Harvard University Press, 1965) and L. W. Gysel, "Borders and Openings of Beech-Maple Woodlands in Southern Michigan," *Journal of Forestry* 49 (1951): 13–19.

19. Australian ecologists like C. R. Margules developed a critique of SLOSS in the 1970s, but did not make a significant impact in the U.S. debates until much later.

20. Francesco Di Castri, F. W. G. Baker, and Malcolm Hadley, eds. *Ecology in Practice* (Paris: UNESCO, 1984), 1:405.

21. Grumbine, *Ghost Bears,* 30.

22. Thomas Lovejoy's work in the Amazon, the Minimum Critical Size of Ecosystems Project, did attempt to test this experiment, but his project was not begun until 1980 and results did not begin to trickle out until much later. The project is still continuing. Lovejoy received his Ph.D. in ecology from Yale University, studying under Evelyn Hutchinson, when MacArthur was one of the young faculty members.

23. Grumbine, *Ghost Bears,* 31.

24. Di Castri, Baker, and Hadley, *Ecology in Practice,* 1:404–5.

25. One of the key early studies here, though it did not crystallize the field like MacArthur and Wilson's "Equilibrium Theory," was R. J. Berry, "Conservation Aspects of the Genetical Constitution of Populations," in *The Scientific Management of Animal and Plant Communities for Conservation,* ed. E. D. Duffey and A. S. Watt (Oxford: Blackwell, 1971), 177–206. Michael Soule is commonly recognized as one of the key advocates for this theory now, since his publication, with B. A. Wilcox, of *Conservation Biology: An Evolutionary-Ecological Perspective* (Sunderland, Mass.: Sinauer Press, 1980).

26. When populations reach a certain low point, they are especially vulnerable to random events, such as a catastrophic fire or disease, or even a randomly skewed sex ratio for a few successive generations. There are also the unforeseeable effects of low populations: passenger pigeons did not become extinct because hunters shot the very last ones out of the sky, for instance. It appears they went extinct because they had behavioral triggers for breeding that depended on large flocks of birds, and once hunters had reduced their population to a certain level their breeding success plummeted. Paul Ehrlich and Ann Ehrlich, *Extinction: The Causes and Consequences of the Disappearance of Species* (New York: Random House, 1981).

27. Michael Soule and Daniel Simberloff, "What Do Genetics and Ecology Tell Us about the Design of Nature Reserves?" *Biological Conservation* 35 (1986): 32.

28. Ibid.

29. Mark L. Shaffer, "Minimum Population Sizes for Species Conservation," *Bio-Science* 31, no. 2 (February 1981): 133.

30. Soule and Simberloff, "Genetics and Ecology," 35.

31. Quammen, *Song of the Dodo,* 484–85.

32. Berwick and Saharia, *Development of Wildlife Research,* 480.

33. Daniel Simberloff and L. G. Abele, "Island Biogeography and Conservation Strategy and Limitations," *Science* 193 (1976): 1032.

34. Soule and Simberloff, "Genetics and Ecology," 23.

35. Quammen, *Song of the Dodo,* 480.

36. For these smaller and less mobile organisms a separation of even a few kilometers is often sufficient to move from one closely related species to another. Thus, to preserve small-scale biodiversity it is advantageous to spread out reserves as much as possible. Terry Erwin, in studies of insects in the Amazon rain forest canopy, has demonstrated that one tree can contain hundreds of species, often differing from the species found in another tree in the same forest. Erwin, "Tropical Forest Canopies."

37. Di Castri, Baker, and Hadley, *Ecology in Practice,* 1:414.

38. Thomas E. Lovejoy and David C. Oren, "The Minimum Critical Size of Ecosystems," in *Forest Island Dynamics in Man-Dominated Landscapes,* ed. Robert L. Burgess and David M. Sharpe, vol. 41 of *Ecological Studies* (New York: Springer Verlag, 1981). This project is described narratively in Quammen, *Song of the Dodo,* 451–98.

39. Soule and Simberloff, "Genetics and Ecology," 25.

40. See chapter 3 for an extended discussion of this cultural attitude as it has manifested itself among U.S. ecologists.

41. Soule and Simberloff, "Genetics and Ecology," 22.

42. Eric Schmitt, "Everglades Restoration Plan Passes House with Final Approval Seen," *New York Times,* October 20, 2000, <http://www.nytimes.com/2000/10/20/science/20GLAD.html>.

43. Michael Soule, introduction to Grumbine, *Ghost Bears,* xi; John MacKinnon and Kathy MacKinnon, eds., *Review of the Protected Areas System in the Indo-Malayan Realm* (Gland, Switzerland: IUCN, 1986), 9; Grumbine, *Ghost Bears,* 56, quoting Shugart and West; Michael Bean, Sarah Fitzgerald, and Michael O'Connell, *Reconciling Conflicts under the Endangered Species Act: The Habitat Conservation Planning Experience* (Washington, D.C.: WWF, 1991), 24; IUCN, *World Conservation Strategy* (Gland, Switzerland: IUCN, 1980); L. G. Abele and Edward Connor, "Application of Island Biogeography Theory to Refuge Design: Making the Right Decision for the Wrong Reasons," in *Proceedings of the First Conference on Scientific Research in the National Parks,* ed. R. M. Linn, National Park Service Transactions and Proceedings Series, no. 5 (Washington, D.C.: U.S. Department of the Interior, 1979), 1:94.

44. Cf. my interviews with ecologists in the United States and in India. For a comprehensive and insightful analysis of the border between ecology and environmentalism, and the role of environmental values in ecological science, see David Takacs, *The Idea of Biodiversity: Philosophies of Paradise* (Baltimore: Johns Hopkins University Press, 1996).

45. Wilson, *Naturalist,* 323. See also E. O. Wilson, *The Insect Societies* (Cambridge, Mass.: Belknap Press of Harvard University Press, 1971) and Wilson, *Sociobiology: The New Synthesis* (Cambridge, Mass.: Belknap Press of Harvard University Press, 1975).

46. On the Wilderness Act, see Nash, *Wilderness,* 220–26. For U.S. environmentalism, see Kirkpatrick Sale, *The Green Revolution: The American Environmental Movement, 1962–1992* (New York: Hill and Wang, 1993).

47. This organization, a nongovernmental alliance of national academies of science from around the world, was founded in 1931.

48. Kwa, "Representations of Nature."

49. E. B. Worthington, ed., *The Evolution of IBP* (Cambridge: Cambridge University Press, 1975), 3–6.

50. Kwa, "Representations of Nature," 415. From 1970 to 1974, $40 million was spent in the United States on the IBP.

51. UNESCO, *Intergovernmental Conference of Experts on the Scientific Basis for Rational Use and Conservation of the Resources of the Biosphere, Final Report: SC/MD/9* (Paris: UNESCO, January 6, 1969), 33. The name of this conference was sensibly shortened in most articles to the Biosphere Conference.

52. Richard Forster, *Planning for Man and Nature in National Parks: Reconciling Perpetuation and Use* (Morges, Switzerland: IUCN, 1973), 26.

53. Takacs, *Idea of Biodiversity*, 4; Wilson, *Naturalist*, 356; Norman Myers, *The Sinking Ark: A New Look at the Problem of Disappearing Species* (New York: Pergamon Press, 1979); Wilson, *Naturalist*, 357.

54. Wilson, *Naturalist*, 357; U.S., Department of State, *Humid Tropical Forests: AID Policy and Program Guidance* (Washington, D.C.: Department of State Memorandum, 1985); Wilson, *Naturalist*, 357.

55. All data on comparative extinction rates are derived from W. V. Reid, "How Many Species Will There Be?" in *Tropical Deforestation and Species Extinction,* ed. T. C. Whitmore and J. A. Sayer (London: Chapman and Hall, 1992), 55–73.

56. S. L. Pimm, G. J. Russell, J. L. Gittleman, and T. M. Brooks, "The Future of Biodiversity," *Science* 269 (1995): 347–50.

57. Takacs, *Idea of Biodiversity*, 332.

58. Wilson, *Naturalist*, 357; Takacs, *Idea of Biodiversity*, 35–39.

59. Takacs, *Idea of Biodiversity*, 37.

60. Ibid., 38. Also, E. O. Wilson, ed., *BioDiversity: National Forum on BioDiversity* (Washington, D.C.: National Academy Press, 1986). Thirteen years after the conference, its promotional poster, a fanciful collage of rain forest species on a lush background, still hangs in the lab of one University of Iowa ecology professor.

61. Michael Soule, "What Is Conservation Biology?" *BioScience* 35, no. 11 (December 1985): 727.

62. Conservation biology, in the first decade of its formal existence, has been primarily oriented to the theories of island biogeography and minimum viable populations, both of which developed in ecology in the sixties and onward. Island biogeography is considered to be community ecology—looking at the interactions between many species—while minimum viable populations is a product of population ecology, oriented to the success of one species population. Island biogeography draws heavily on the mathematical and theoretical revolution within ecology, while minimum viable population theories are heavily dependent on genetics. Conservation biology in its modern incarnation is thus often perceived as a strictly post-1965 science. It also draws from and incorporates much earlier ecological practices, however, including the behavioral interaction studies of Ali and the early ecologists in the United States and Europe.

For more, see Daniel Simberloff, "The Contribution of Population and Community Biology to Conservation Science," *Annual Review of Ecology and Systematics* 19 (1988): 473–511.

63. Soule, "What Is Conservation Biology?" 733.

64. Takacs, *Idea of Biodiversity,* 4.

65. Denis Saunders, Richard Hobbs, and Chris Margules, "Biological Consequences of Ecosystem Fragmentation: A Review," *Conservation Biology* 5, no. 1 (March 1991): 19–20.

66. Maslow, *Footsteps in the Jungle,* 261–88.

Chapter 7

1. The WII was granted autonomy in 1986, the same year as the invention of the term *biodiversity* and the formation of the U.S.-based Society for Conservation Biology.

2. Among his many activities, Ranjitsinh conceptualized and wrote the 1972 Wildlife (Protection) Act that established the Indian national park system, and he was undersecretary of the Project Tiger committee.

3. Interview with a retired member of the Indian Administrative Service who has chosen to remain anonymous, WWF-India, New Delhi, July 13, 1998.

4. Barro Colorado Island, in Panama, run by the Smithsonian Institution, is for all intents a U.S. property in Central America.

5. For an example see McClure, *Stories,* 278.

6. Interview with a scientist who has chosen to remain anonymous, Iowa Lakeside Laboratory, July 22, 1997.

7. This is similar to Visvanathan's argument, explained in chapter 1, that to conduct Western science is to inherently be Westernized and to be oriented to Western ways of seeing the world.

8. Zafar Futehally to Lee Talbot, March 12, 1970, SIA, RU 271, box 14, f. "Futehally."

9. Juan Spillett to Wayne Mills, March 23, 1972; Spillett to Kennedy B. Schmertz, November 30, 1972, memorandum, SIA, RU 254, box 99, f. 14.

10. Ranjitsinh to Thomas Foose, September 13, 1973, SIA, RU 254, box 100, f. 2.

11. Ripley to Fakraddin Ali Ahmad, April 21, 1972, SIA, RU 254, box 99, f. 14.

12. Memorandum: For the Record, from Ross Simons, May 22, 1974, SIA, RU 254, box 100, f. 3.

13. Almost none of the members of the IFS were trained in ecology—entrance was based on their score on a national test, and in fact most foresters were trained as engineers. They did take a short forestry course upon entering the IFS, but it was not focused on wildlife.

14. Ripley had made an address to the BNHS members in May 1967 that (with accompanying press coverage) he thought had started the ball rolling toward action for the tiger. Things really moved forward at the 1969 IUCN meeting in New Delhi as

several people encouraged tiger conservation—a meeting for which Sankhala was the organizing secretary and at which he presented a paper, "The Vanishing Indian Tiger." In truth, both men contributed to the publicizing of the issue, but neither was first; both were following in the footsteps of Schaller and Gee. Memorandum: For the Record, from Ross Simons, May 22, 1974, SIA, RU 254, box 100, f. 3, also, Ripley to Ambassador Daniel Moynihan, January 11, 1973, "Aide Memoire," SIA, RU 254. The Assistant Secretary for Science, 1963–1978. box 100, f. 1. For a detailed discussion of Sankhala's perspective, see Kailash Sankhala, *Tiger! The Story of the Indian Tiger* (New York: Simon and Schuster, 1977).

15. Ripley to Moynihan, January 11, 1973, "Aide Memoire," SIA, RU 254, box 100, f. 1.

16. Sankhala, *Tiger!* 194.

17. Ripley to Moynihan, January 11, 1973.

18. John Seidensticker, "Tiger Research Development Trip," March, 1972, SIA, RU 271, box 15, f. 2.

19. Seidensticker to Mills, October 13, 1972, SIA, RU 254, box 15.

20. Futehally to Karan Singh, July 22, 1973, SIA, RU 254, box 100, f. 2.

21. Ripley to Guy Mountfort, April 5, 1973, SIA, RU 254, box 100, f. 1.

22. David Challinor to Ripley, November 3, 1972, memorandum, SIA, RU 254, box 24, f. 4.

23. For a description of Sunquist's work given by his wife, see Fiona Sunquist, *Tiger Moon* (Chicago: University of Chicago Press, 1988).

24. Memorandum: For the Record, from Ross Simons, May 22, 1974, SIA, RU 254, box 100, f. 3.

25. An exhaustive array of telegrams, letters, and press clippings can be found in SIA, RU 254, box 100, f. 3.

26. "Tiger, Tiger, Burning Bright, How Did Thou Die?" *National Herald,* September 5, 1974, 3. Most tiger scientists today consider tigers to be territorial.

27. Sankhala reiterated this claim in *Tiger!* 39.

28. Wayne Mills to David Challinor, September 17, 1974, memorandum, SIA, RU 254, box 100, f. 3.

29. The house journal of the IFS, the *Indian Forester,* contains many of their observations, as does the *JBNHS* in its "Notes" section.

30. Mahesh Rangarajan, "The Politics of Ecology: The Debate on Wildlife and People in India, 1970–95," *Economic and Political Weekly,* September, 1996, 2403. Tour companies were still selling "tiger hunts" into 1969 in India.

31. One example was a preoccupation with water holes. Because water holes attract wildlife (useful when placed near a lookout on which a hunter could wait), their creation became a standard practice in managing a forest for game animals. However, some foresters continued to advocate water hole construction long after the original reason for placing them has passed. One biologist at the WII claimed, "Everybody was still looking to create water holes." He went on to argue that water holes create internal

edges within forests and favor some species, such as chital, over others. They are not universally good for wildlife—they are good for some species, and good for hunting.

32. The history of wildlife training within the FRI and WII is given in V. B. Saharia, "The Building Blocks of Wildlife Education in India," *WII Newsletter*, October 1992 (commemorative issue, 1982–91), 7–9.

33. Indian Board for Wild Life, *Task Force: Project Tiger: A Planning Proposal for Preservation of Tigers (Panthera tigris tigris Linn.) in India* (New Delhi: Ministry of Agriculture, 1972).

34. I could not find the attendance for the 1974–75 diploma course. It seems likely that with Nair's death the course was disbanded. The second course (1979–80) trained only ten foresters; four subsequent courses trained: 18, 19, 18, and then 30, foresters, respectively. The last two courses were given after the WII had been formed. WII, *Wildlife Institute of India: A Profile* (Dehradun: Wildlife Institute of India, 1993), Annexure [Appendix] 1.

35. Valmik Thapar, "Fatal Links," *Seminar* 466 (June 1998): 64.

36. I have been unable to find any scientific publications in refereed journals on tigers in India from the seventies to the mid-eighties

37. Interview with Ullas Karanth, IISC, Bangalore, July 7, 1998.

38. It was easier in the early seventies for European scientists to gain access to India than it was for U.S. scientists—Martin was one of the few who took advantage of this.

39. Interview with H. Panwar, March 6, 1998.

40. The FAO, a branch of the U.N., includes many environmental programs in its purview, despite its seemingly agricultural focus. The FAO often works in conjunction with the UNDP, and the two organizations are joint sponsors of the Man and the Biosphere program (with biosphere reserves). The FAO has a fisheries and a forestry division, and both divisions work with environmental management.

41. In the last thirty years the FAO has supported crocodile research throughout the developing world, including Africa, Latin America, and Asia.

42. There is a very vocal group of crocodile experts in India who argue that crocodiles should be farmed for their skins and the proceeds used to support conservation measures for other species. See, for example, Indraneil Das, "Animal Farm," *Seminar* 466 (June 1998): 50–52. The three crocodile species are still endangered in the wild (although they will never go extinct due to the success of captive breeding), so the sale of crocodile skins is illegal in India.

43. General information on the history of Project Crocodile was derived from interviews with scientists who participated in the project, from J. C. Daniels, and from the *WII Newsletter*, October 1992.

44. David Challinor to Ripley, memorandum, October 10, 1979, SIA, RU 7008, acc. 92–063, box 1.

45. Ali to Ripley, November 23, 1979, SIA, RU 7008, acc. 92–063, box 1.

46. Berwick and Saharia, *Development of Wildlife Research*, v. Note the difference—

the United States contributed scientists, India contributed forestry officials and scientists. In 1982, Saharia would have had no problem finding fifteen Indian scientists, had he wished to do so, at the BNHS and CES alone. As director of the forestry wildlife division, though, Saharia chose to anoint some of the IFS naturalists as the Indian experts for this conference. Saharia, "Wildlife Education in India," 8.

47. Note also that this workshop was oriented toward observational wildlife ecology, not theoretical ecology, or even botany, for that matter. Saharia, "Wildlife Education in India," 8; Berwick and Saharia, *Development of Wildlife Research,* v.

48. Saharia, "Wildlife Education in India," 9.

49. Ibid., 8; M. K. Ranjitsinh, Joint Secretary to the Government of India, "Autonomy Document for Wildlife Institute of India," Internal memo no. 23–17/84-FRY (WL), New Delhi: Department of Environment, Forests, and Wildlife (Wild Life Division), 1986.

50. The same nine- and three-month courses that had been offered before—the ten and a half month course had been shortened slightly.

51. Interview with A. J. T. Johnsingh. All anecdotal information is derived from the Johnsingh interviews that took place at the WII, between March 5 and April 3, 1998.

52. Michael Fox to H. Buechner, April 10, 1973, SIA, RU 274, box 9, f. 7.

53. Fox was also unusual in coauthoring an article with Johnsingh for the *JBNHS,* rather than just writing one himself. Michael Fox and A. J. T. Johnsingh, "Hunting and Feeding in Wild Dogs," *JBNHS* 72, no. 2 (1974): 321–26.

54. Now called Madhurai Kamaraj University.

55. Although many people told me that Johnsingh was the first Indian wildlife biologist with a Ph.D., his work was actually contemporary with the Ph.D. research of the people in Project Crocodile. If you consider only wildlife biologists who work on wild mammals, Johnsingh does appear to be the first, preceding Sukumar's elephant work at the IISc by a couple of years.

56. WII, *Wildlife Institute of India: Annual Report 1987–88* (New Forest, Dehradun: Wildlife Institute of India, December 1988), 23.

57. WII, *Annual Report, 1987–88,* 23–24. In 1987—88, for example, one scientist went to Colorado State University for six months to study habitat evaluation; another went to Australia for six weeks to study tropical grassland management.

58. The FAO and the UNDP have continued to support the WII, but on a small scale and without on-site experts.

59. WII, *Annual Report, 1987–88,* 31.

60. Interview with Ravi Chellam. The interviews with Chellam occurred at the WII between March 2 and March 17, 1998.

61. WII, *Annual Report, 1986–87,* 13–16.

62. Interview with S. P. Goyal, WII, March 25, 1998.

63. James Teer to B. R. Seshachar, May 15, 1970, SIA, RU 271, box 14.

64. Interview with a former administrator at WII who has chosen to remain anonymous, WII, March 6, 1998.

65. Interview with a biologist who has chosen to remain anonymous, New Delhi, May 4, 1998.

66. WII, *Annual Report 1986–87,* 5.

67. Budiansky, *Nature's Keepers.*

68. Interview with Y. V. Jhala, WII, March 19, 1998; interview with a U.S. ecologist who has chosen to remain anonymous, May 6, 1999.

69. Ajith Kumar and Ravi Chellam, "Project Proposal: Impact of Fragmentation on the Biological Diversity of Rain Forest Small Mammals and Herpetofauna of the Western Ghats Mountains, South India," manuscript, 1991, 8.

70. Chellam interview.

71. Most of the small mammals she studied cannot be found in zoos, so she was unable to obtain comparative fecal samples from that source. The exception is the brown palm civet—there is one of these in the Pune zoo—but it is fed mice, which they do not eat in the wild, so it is not of much use as a fecal "type" specimen.

72. According to Ravi Chellam, who proudly showed me some of these pictures in his office. In S. H. Prater's definitive *Book of Indian Animals* there are not even drawings of two of her three primary study species.

73. Interview with G. S. Rawat, WII, March 4, 1998.

74. Saharai and Berwick, *Development of Wildlife Research,* v–vi. Also WII, *Annual Report, 1986–87,* 5.

75. Stories of local assimilation of foreign technologies are common—in Nepal I saw a toilet bowl, with holes in the bottom, turned upside down so that the roots of vegetables growing through the holes would be protected from animals.

76. Material in the last two paragraphs is from an interview with S. A. Hussain, WII, March 25, 1998.

77. As has been evident throughout this book, historically, privilege has been almost a precondition for success in Indian ecology. Ali and Gadgil both grew up in wealthy families who supported their sons' eccentric pursuit of ecology studies. This is changing, slightly. Though most wildlife biologists at the WII have come from economically comfortable families, there are exceptions who gain entry to the institute by virtue of their performance on the standardized entry examinations. Success on these examinations for people who did not attend English-medium schools (often private schools) is very difficult, however. So while the examination is ostensibly meritocratic, the deck is stacked against students who have not had access to top-level secondary and undergraduate education.

78. As mentioned in chapter 1, the government of India was very resistant to radio telemetry for wildlife biology but not for these reasons. At various times it was feared that the anesthesia to capture the animal would possibly kill an endangered species, that the collar itself was harmful, and, on at least one occasion, that the CIA could use positional data from the transmitter that would compromise Indian national security.

79. Amita Baviskar, "The Community and Conservation: The Case of Ecodevelopment in the Great Himalaya National Park, India," manuscript given to me by the

author, 1997, 8. GHNP is a mountainous park of 765 square kilometers in Kulu District of Himachal Pradesh. Most of the park is above four thousand meters and three ridges reach above fifty-five hundred meters. It is a relatively new park, its notification as a national park occurring in 1984, primarily at the recommendation of two biologists, a British and a Canadian, who surveyed the region and declared it both worthy of and in need of preservation.

80. Vinod follows a naming practice found in Kerala, but uncommon elsewhere, in which the caste name is obscured by using only an initial. Much more common in India is the reverse, in which the given name is presented as an initial and the family name is used (Y. V. Jhala, for instance).

Chapter 8

1. Hundreds of thousands may be a low estimate—there are two hundred thousand people who live within a few kilometers of Rajaji National Park alone.

2. P. J. Dugan, ed., *Wetland Conservation: A Review of Current Issues and Required Action* (Gland, Switzerland: IUCN/ World Conservation Union, 1990).

3. Bishnu Bhandari, "Institutions and Capability in Wetland Management in Nepal," in *Safeguarding Wetlands in Nepal: Proceedings of the National Workshop on Wetlands Management in Nepal,* ed. Bishnu Bhandari, T. B. Shrestha, and John McEachern (Gland, Switzerland: IUCN/World Conservation Union, 1994), 121.

4. This has been especially true in the United States. States such as Iowa and California have converted over 95 percent of their wetlands into agricultural land.

5. Martin Ewans, *Bharatpur: Bird Paradise* (London: H. F. and G. Witherby, 1989), 9.

6. Schaller, *Deer and Tiger,* 4–7.

7. Ali, *Fall of a Sparrow,* 150.

8. Ibid., 150; Mahesh Rangarajan, *India's Wildlife History* (Delhi: Permanent Black, 2001), 82–85.

9. Ali, *Fall of a Sparrow,* 150.

10. Mahesh Rangarajan, *Troubled Legacy,* 18; Ali, *Fall of a Sparrow,* 150–51.

11. Ewans, *Bharatpur,* 10.

12. Rangarajan, "Politics of Ecology," 2406.

13. Sankhala, "Livestock Grazing," 58; E. P. Gee, "The Management of India's Wild Life Sanctuaries and National Parks," *JBNHS* 64, no. 2 (1967): 339. When I spoke with a senior scientist in the BNHS, I asked if it was reasonable to have pristine core areas in the national parks where people and livestock were excluded. Do the cattle and villagers also have some right to the land? His response: "Certainly not the cattle. Billions of them. Miserable animals, and they should be——." He never finished his statement—an aide walked in and we changed topics when we resumed.

14. Schaller, *Deer and Tiger,* 6–7.

15. J. Juan Spillett, "General Wild Life Conservation Problems in India," *JBNHS* 63, no. 3 (1966): 617.

16. Ibid., 617, 620.

17. Vasant Saberwal, *Pastoral Politics: Shepherds, Bureaucrats, and Conservation, in the Western Himalaya* (Delhi: Oxford University Press, 1999).

18. Spillett, "Conservation Problems," 618.

19. Editorial, *JBNHS* 63, no. 3 (1966): 491.

20. Bittu Sahgal, "A Birdwatcher's Eye-View," *Express Magazine,* September 11, 1983, 5.

21. Peter Denton, *Organizations That Help the World: World Wide Fund for Nature* (New Delhi: Orient Longman, 1996), 16.

22. B. R. Sesachar to A. K. Sarkar, October 22, 1967, SIA, RU 271, box 14, f. 1.

23. Zafar Futehally, "Misuse of Nature: Some Ecological Facts," *Times of India* June 11, 1967.

24. B. R. Seshachar to Zafar Futehally, October 24, 1967, SIA, RU 271, box 14, f. 1.

25. Chester Bowles to Ripley, November 7, 1967, SIA, RU 271, box 14, f. 1.

26. I did not find Joslin's paper, but I do have a copy of the report on which it was based: the report from the 1969 IUCN conference in New Delhi, which proved critical in the establishment of Project Tiger. Stephen Berwick and Peter Jordan, "Conservation of a Natural Ecosystem in the Gir Forest, India," unpublished report, SIA, RU 274, box 9, f. 7.

27. Talbot to Paul Joslin, November 11, 1969, SIA, RU 271, box 14, f. 4.

28. Jenkins to Paul Joslin, December 8, 1969, SIA, RU 271, box 14, f. 4.

29. When I was at Corbett National Park, I was thrilled to see a red junglefowl, a beautiful bird, running through the forest. The junglefowl is the direct ancestor of the chicken. In fact, it looks just like a chicken. But it was elusive, it was moving fast, it is listed in *The Book of Indian Birds,* and I was excited. In contrast, when I saw cattle in a national park I was always disappointed.

30. A bund is an embankment used to direct water. Ripley to Ali, February 11, 1980, SIA, RU 7008, acc. 92–063, box 1.

31. N. D. Jayal, "Note on a Tour of Keoladeo Ghana Bird Sanctuary at Bharatpur," unpublished report, July 1976, SIA, RU 254, box 24, f. 3.

32. Ripley and Ali corresponded about the event—they were concerned that the ban on cattle had been poorly implemented. Ripley had written a draft letter to Gandhi urging her to hold firm on the ban, but Ali suggested he not mail it. Ripley to Ali, December 6, 1982; Ali to Ripley, December 16, 1982; SIA, RU 7008, acc. 92–063, box 1.

33. V. S. Vijayan, *Keoladeo National Park Ecology Study* (Bombay: Bombay Natural History Society, 1987).

34. V. S. Vijayan, *Keoladeo National Park Ecology Study 1980–1990 Summary* (Bombay: Bombay Natural History Society, 1991), 19.

35. Ibid., 1.

36. Budiansky, *Nature's Keepers,* 142.

37. Nathaniel P. Reed, "How Well Has the United States Managed Its National Park System? The Application of Ecological Principles to Park Management," in *Second*

World Conference on National Parks: Yellowstone and Grand Teton National Parks, USA, September 18–27, 1972, ed. Hugh Elliott (Morges, Switzerland: International Union for Conservation of Nature and Natural Resources, 1974), 40.

38. Budiansky offers a spirited critique of this move and documents in detail the political machinations that resulted in this policy, including thousands of letters from school children decrying the "culling" of elk and protests from big-game hunters in Montana who needed animal overpopulations in Yellowstone to supply them with animals to hunt in bordering lands when the animals wandered out of the park. Budiansky, *Nature's Keepers,* 131–55.

39. In 1962 the First World Conference on National Parks was held in Seattle, sponsored by the U.S. National Park Service, the IUCN, UNESCO, and the FAO. There were 145 international delegates, and 117 from the United States (who, though the panels were often international, dominated many of the "commentator" and "summary" positions, as well as provided the keynote address). The stated purpose of the meeting was to "establish a more effective international understanding and to encourage the national park movement on a worldwide basis." "How It Began," in *First World Conference on National Parks, Seattle, Washington, June 30–July 7, 1962,* ed. Alexander Adams (Washington, D.C.: National Park Service, U.S. Department of the Interior, 1962), xxxii.

Ten years later there was a second such meeting, held at Yellowstone National Park in honor of its centennial. Again, the purpose of this conference was to spread the National Park gospel, and this time, the gospel included Yellowstone's new policy of "natural regulation." Reed, "How Has the U.S. Managed?" 40. An explicit claim for the U.S. National Park System as a global model is made by Jean-Paul Harroy of Belgium in "A Century in the Growth of the 'National Park' Concept throughout the World," in *Second World Conference on National Parks: Yellowstone and Grand Teton National Parks, USA, September 18–27, 1972,* ed. Hugh Elliott (Morges, Switzerland: International Union for Conservation of Nature and Natural Resources, 1974), 24–32.

40. See Mahesh Rangarajan, *Fencing the Forest: Conservation and Ecological Change in India's Central Provinces, 1860–1914* (Delhi: Oxford University Press, 1996).

41. Charles E. Kay, "Yellowstone's Northern Elk Herd: A Critical Evaluation of the 'Natural Regulation' Paradigm" (Ph.D. diss., Utah State University, 1990).

42. Budiansky, *Nature's Keepers,* 141.

43. Interview with Shruti Sharma, Bharatpur, December 28, 1997.

44. While I was at Bharatpur, I saw a constant stream of villagers with headloads of grass leaving the park.

45. Budiansky, *Nature's Keepers,* 152–53.

46. One place where this is occurring is the Valley of Flowers National Park, whose centerpiece is a valley meadow of astonishing floral diversity. Traditionally, high Himalayan meadows served as summer grazing areas for nomadic herders. Researchers from the WII claim that banning herding is helping the park; researchers from Delhi University claim that grazing has helped control a pesky weed that would

otherwise have taken over the valley much earlier. These studies have not been published and it is difficult to adjudicate between their different claims at this point. Informal interviews with researchers at Delhi University and the WII, February, April 1998.

47. Madhav Gadgil, *Establishment of Biosphere Reserves in India: Project Document 1: The Nilgiri Biosphere Reserve* (New Delhi: Indian National Man and the Biosphere Committee, Department of Environment, Government of India, 1980).

48. All dates are derived from Ministry of Environment and Forests, *Biosphere Reserves in India* (New Delhi: Ministry of Environment and Forests, 1989), 23–25. Anecdotal evidence concerning Gadgil's experiences are derived from a series of interviews with Gadgil, conducted in June and July 1998.

49. UNESCO, *Intergovernmental Conference*, 33.

50. To this day the program is still referred to as *Man* and the Biosphere. There has been no move toward gender inclusion, and this holds true in the other languages used by the MAB program (Spanish and French).

51. Gadgil, *Project Document 1*, vi, 7.

52. Cyrille de Klemm. *Biological Diversity Conservation and the Law: Legal Mechanisms for Conserving Species and Ecosystems* (Gland, Switzerland: IUCN, 1993), 3.

53. William Baker, "The Landscape Ecology of Large Disturbances in the Design and Management of Nature Reserves," in *Environmental Policy and Biodiversity*, ed. Edward Grumbine (Washington, D.C.: Island Press, 1994), 84.

54. F. Di Castri, "The Programme on Man and the Biosphere (MAB)," in *Proceedings of the First International Congress of Ecology: Structure, Functioning, and Management of Ecosystems* (Wageningen, Netherlands: Centre for Agricultural Publishing and Documentation, 1974), 480.

55. The parks selected were Everglades National Park, Great Smoky Mountains National Park, and Olympic National Park. At least at Everglades National Park there is a small kiosk at one of its visitor centers that announces that it is a biosphere reserve.

56. Jeffrey McNeely and David Pitt, eds., *Culture and Conservation: The Human Dimension in Environmental Planning* (Dover, N.H.: Croom Helm for IUCN, 1985), 41.

57. Many governments, particularly in the developed world, "considered biosphere reserve designations more as an honor than as an obligation," and most biosphere reserves never acquired the hoped-for core/buffer design. International Biosphere Reserve Congress, *Conservation, Science, and Society: Contributions to the First International Biosphere Reserve Congress* (Minsk: UNESCO/UNEP, September 26–October 2, 1983), introduction.

58. For instance, Conservation International, a U.S. NGO, has established two multizone biosphere reserves thus far in Latin America, one in collaboration with the government of Costa Rica, the other in a debt-for-nature swap with the government of Bolivia in 1986.

59. Interview with Raman Sukumar, IISc, Bangalore, June 12, 1998.

60. This ministry had been recently formed, after a long existence as the Department of the Environment under the Ministry of Agriculture.

61. Interview with R. Sukumar.

62. Specific accounts of funds disbursed in the first three years of the biosphere reserve program are given in Ministry of Environment and Forests, *Biosphere Reserve in India,* 19. In its first three years of formal existence, the Nilgiri Biosphere Reserve received 1.16 million rupees. It is very difficult to trace more recent funding within the biosphere reserve program or to know exactly where this money went, as government records are closed. It is possible that some of this money was double-reported; in other words, it was money given to government ministries working in the Nilgiri area, who would have received funding in any case.

63. Michael Godt, public address at the Biodiversity Preservation Prioritization Conference, WWF-India office, New Delhi, April 26, 1998.

64. This forest officer, who was in attendance at the BCCP conference, prefers to remain anonymous, April 26, 1998.

65. In fact, I heard this story nowhere outside Bangalore.

66. Gadgil, public address on biosphere reserves at the BCCP conference, New Delhi, April 26, 1998.

67. Indira Gandhi took several unilateral actions as prime minister based on her environmental concern, the most famous being her intervention in a hydroelectric project that threatened to destroy a pristine rain forest in southern India called Silent Valley.

68. This person has chosen to remain anonymous; his interview was conducted on January 16, 1998.

69. Evans, *Bharatpur.*

70. MacKinnon and MacKinnon, *Protected Areas System.*

71. Bharatpur, at approximately four hundred fifty hectares, would not qualify as a national park under the IUCN's criteria of size discussed in chapter 6, even with the more liberal 1967 requirements, which allowed heavily populated countries to have national parks as small as five hundred hectares.

72. Nomination form, Biosphere Reserves, MAB Program, UNESCO; emphasis in original.

73. Madhav Gadgil and Ram Guha, *This Fissured Land: An Ecological History of India* (Delhi: Oxford University Press, 1992); *Ecology and Equity.* Although Gadgil is to be commended for crossing this disciplinary boundary, some of his social science work has met with mixed reactions, particularly his dubious explanation of the caste system.

74. Gadgil and Vartak, "Sacred Groves."

75. Ibid., 313.

76. Guha, "Authoritarian Biologist," 16, 19.

77. Visvanathan, *Carnival for Science,* 92.

78. This reiterates the position taken by the government of India regarding

Bharatpur in the fifties, when Ali approached Nehru to stop plans to convert the wetlands into agricultural land.

79. Singh, *Biodiversity Conservation,* 29. This publication, which was a version of
the grant application, was published by UNESCO's MAB program.

Bibliography

A Note on Sources

Much of the research for this volume did not involve traditional archival research. As my topic involves recent history, the records in the archives of the government of India are closed. Some current and former government bureaucrats allowed me to make copies of reports and memorandums from their private collection, but such access was sporadic at best (and in one case extensively censored). Thus, for government-sponsored programs like biosphere reserves and national park policies, as well as the establishment of the CES and the WII, I was at times forced to rely on whatever records were publicly available in the libraries of these institutions. Fortunately I was able to fill in many blanks with the Smithsonian Institution Archives (SIA), an underutilized resource for South Asian environmental history. The connections between leading Indian figures (both scientists and government officials) and the Smithsonian are wide ranging, and the SIA cover ground far beyond environmental topics. The family of S. Dillon Ripley allowed me to use the Ripley collection at the Smithsonian, which is restricted until 2011 and is still in the process of being catalogued. This project would have been much weaker without access to this collection.

In the absence of traditional archival sources, I used interviews to derive much of my information about the practice of ecology, as well as environmental policies and politics, in the last fifty years. Throughout India and the United States I found ecologists, government officials, and environmentalists to be extraordinarily helpful. In addition to these formal interviews, I was also assisted in my research by conversations with graduate students, scientists, environmentalists, forest officers, and nature guides in both countries.

Smithsonian Institution Archives

Record Unit 134. The Canal Zone Biological Area, 1918–64.
Record Unit 245. The Pacific Ocean Biological Survey Program, National Museum of Natural History, circa 1961–73, with data from 1923.

Record Unit 254. The Assistant Secretary for Science, 1963–78.

Record Unit 271. Ecology Program, Office of Environmental Sciences, 1965–73.

Record Unit 274. Office of International Programs, 1964–76.

Record Unit 7008. S. Dillon Ripley Papers, 1913–93 and undated, with related materials from 1807, 1871–91.

Record Unit 9591. S. Dillon Ripley Interviews, 1977–83.

Books and Articles

Abbot, C. G. "The Smithsonian Institution–Its Activities and Capacities." In *Proceedings of the Conference on the Future of the Smithsonian Institution, February 11, 1927*. Baltimore: Lord Baltimore Press, 1927.

Abele, L. G., and Edward Connor. "Application of Island Biogeography Theory to Refuge Design: Making the Right Decision for the Wrong Reasons." In *Proceedings of the First Conference on Scientific Research in the National Parks,* ed. R. M. Linn, 1:89–94. National Park Service Transactions and Proceedings Series, no. 5. Washington, D.C.: U.S. Department of the Interior, 1979.

Abrahamson, Warren G., and Arthur E. Weis. *Evolutionary Ecology across Three Trophic Levels: Goldenrods, Gallmakers, and Natural Enemies.* Princeton: Princeton University Press, 1997.

Adams, Alexander, ed. *First World Conference on National Parks,* Seattle, Washington, June 30–July 7, 1962, Washington, D.C.: U.S. Department of the Interior, 1962.

Agarwal, Anil, and Sunita Narain. *Towards a Green World.* New Delhi: Centre for Science and Environment, 1992.

Agarwal, Anil, et al., eds. *The State of India's Environment, 1982: A Citizen's Report.* New Delhi: Centre for Science and Environment, 1982.

Agarwal, S. K., and R. K. Garg, eds. *Environmental Issues and Researches in India.* Udaipur: Himanshu Publications, 1988.

Ali, Sálim. "The BNHS/WHO Bird Migration Study Project–4." *Journal of the Bombay Natural History Society* 61, no. 1 (March 1964): 99–107.

———. *The Book of Indian Birds.* 12th ed., rev. and enl. Delhi: Oxford University Press, 1997 [1941].

———. *The Fall of a Sparrow.* Delhi: Oxford University Press, 1985.

———. "Flower-Birds and Bird-Flowers in India." *Journal of the Bombay Natural History Society* 34 (1932): 573–605.

———. "The Moghul Emperors of India as Naturalists and Sportsmen." Parts 1–3. *Journal of the Bombay Natural History Society* 31–32 (1927–28).

———. "The Nesting of the Baya (*Ploceus philippinus*): A New Interpretation of Their Domestic Relations." *Journal of the Bombay Natural History Society* 34 (1931): 947–64.

———. "Ornithology in India: Its Past, Present, and Future." *Proceedings of the Indian Science Academy* 33B, no. 3 (1971).

———. Review of *The Wildlife of India* by E. P. Gee. *Journal of the Bombay Natural History Society* 61, no. 2 (1964): 424–26.

Ali, Sálim, and S. Dillon Ripley. *Handbook to the Birds of India and Pakistan.* 10 vols. New Delhi: Oxford University Press, 1968–74.

Alvares, Claude. *Fish, Curry, and Rice: A Citizen's Report on the State of the Goan Environment.* Mapusa: Other India Press, 1993.

Anderson, David, and Richard Grove. *Conservation in Africa: People, Policies, and Practice.* Cambridge: Cambridge University Press, 1987.

Anderson, Robert. *Building Scientific Institutions in India: Saha and Bhabha.* Montreal: McGill University Paper Series no. 11, 1975.

Appadurai, Arjun. *Modernity at Large: Cultural Dimensions of Globalization.* Minneapolis: University of Minnesota Press, 1996.

Arnold, David. *Colonizing the Body: State Medicine and Epidemic Disease in Nineteenth-Century India.* Berkeley: University of California Press, 1993.

Baker, William. "The Landscape Ecology of Large Disturbances in the Design and Management of Nature Reserves." In *Environmental Policy and Biodiversity,* ed. R. Edward Grumbine, 75–98. Washington, D.C.: Island Press, 1994.

Bandhu, Desh, and Eklavya Chauhan, eds. *Current Trends in Indian Environment: Proceedings of the Workshop on Environment Education, Held at the National Science Academy, New Delhi, on July 11–12, 1977.* New Delhi: Today and Tomorrow's Printers and Publishers, 1977.

Banerjee, K. "Emerging Viral Infections with Special Reference to India." *Indian Journal of Medical Research* 103 (April 1996): 177–200.

Barbier, Edward, Joanne Burgess, and Carl Folke. *Paradise Lost: The Ecological Economics of Biodiversity.* London: Earthscan Publications, 1994.

Barbour, Michael. "Ecological Fragmentation in the Fifties." In *Uncommon Ground,* ed. William Cronon, 233–55. New York: Norton, 1995.

Barua, Shankar. "Of Pocket Sanctuaries: A Concept, a Proposal to Monitor a Small Lake near Delhi." Manuscript in WWF-India Library, n.d.

Basella, George. "The Spread of Western Science." *Science* 156 (1967): 611–22.

Baviskar, Amita. "The Community and Conservation: The Case of Ecodevelopment in the Great Himalayan National Park, India." Manuscript, 1997.

Bean, Michael, Sarah Fitzgerald, and Michael O'Connell. *Reconciling Conflicts under the Endangered Species Act: The Habitat Conservation Planning Experience.* Washington, D.C.: World Wildlife Fund, 1991.

Bederman, Gail. *Manliness and Civilization: A Cultural History of Gender and Race in the United States.* Chicago: University of Chicago Press, 1995.

Berry, R. J. "Conservation Aspects of the Genetical Constitution of Populations." In *The Scientific Management of Animal and Plant Communities for Conservation,* ed. E. D. Duffey and A. S. Watt. Oxford: Blackwell, 1971.

Berry, Wendell. "Preserving Wilderness." *Wilderness,* Spring 1987, 39–40, 50–53.

Berwick, Stephen, and V. B. Saharia, eds. *The Development of International Principles and Practices of Wildlife Research and Management: Asian and American Approaches.* Delhi: Oxford University Press, 1995.

Bhandari, Bishnu. "Institutions and Capability in Wetland Management in Nepal." In *Safeguarding Wetlands in Nepal: Proceedings of the National Workshop on Wetlands Management in Nepal,* ed. Bishnu Bhandari, T. B. Shrestha, and John McEachern. Gland, Switzerland: IUCN/World Conservation Union, 1994, 118–35.

Bhandari, Bishnu, T. B. Shrestha, and John McEachern, eds. *Safeguarding Wetlands in*

Nepal: Proceedings of the National Workshop on Wetlands Management in Nepal. Gland, Switzerland: IUCN/World Conservation Union, 1994.

Bhasin, M. K. *Cold Desert, Ladakh: Ecology and Development.* Delhi: Kamla-Raj Enterprises, 1992.

Bhasin, Veena. *Ecology, Culture, and Change: Tribals of Sikkim Himalayas.* New Delhi: Inter-India Publications, 1989.

——. *Himalayan Ecology, Transhumanance, and Social Organisation.* Delhi: Kamla-Raj Enterprises, 1988.

Biswas, Dilip K. *National Committee on Environmental Planning and Coordination: Recommendations and Activities, 1972–81.* New Delhi: Department of Environment, Government of India, 1982.

Blouin, Michael, and Edward Connor. "Is There a Best Shape for Nature Reserves?" *Biological Conservation* 23, 1985, 277–88.

Boffey, Philip M. "Biological Warfare: Is the Smithsonian Really a Cover?" *Science* 163 (1969): 791–96.

——. "Nerve Gas: Dugway Accident Linked to Utah Sheep Kill." *Science* 162 (1968): 1460–64.

Bonner, Raymond. *At the Hand of Man: Peril and Hope for Africa's Wildlife.* New York: Vintage Books, 1994.

Bonta, Marcia Myers. *Women in the Field: America's Pioneering Women Naturalists.* College Station: Texas A&M University Press, 1991.

Brown, Judith M. *Nehru: Profiles in Power.* Harlow, England: Pearson Education, 1999.

Budiansky, Stephen. *Nature's Keepers: The New Science of Nature Management.* New York: Free Press, 1995.

Bush, Mark. *Ecology of a Changing Planet.* Upper Saddle River, N.J.: Prentice-Hall, 1997.

Caldwell, Lynton. *In Defense of the Earth: International Protection of the Biosphere.* Bloomington: Indiana University Press, 1972.

——. *U.S. Interest and the Global Environment.* Occasional Paper no. 35, Stanley Muscatine Foundation, Muscatine, Iowa, 1985.

Callicott, J. Baird, and Michael Nelson, eds. *The Great New Wilderness Debate.* Athens: University of Georgia Press, 1998.

CES (Centre for Ecological Sciences). *A Decade of Ecological Research and Training, 1983–1993.* Bangalore: Centre for Ecological Sciences, IISc, 1993.

——. *Ecological Sciences Research and Training Centre at the Indian Institute of Science, Bangalore: Annual Report for the Period 1–4–90 to 31–3–91.* Bangalore: Centre for Ecological Sciences, IISc, 1991.

Chadha, S. K., ed. *Environmental Crisis in India.* Dehradun: International Book Distributors, 1992.

Chambers, David Wade. "Does Distance Tyrannize Science?" In *International Science and National Scientific Identity,* ed. R. W. Home and S. G. Kohlstedt, 19–38. Boston: Kluwer Academic Publishers, 1991.

Chellam, Ravi, and A. J. T. Johnsingh. "Management of Asiatic Lions in the Gir Forest, India." *Symposia of the Zoological Society of London* 65, 1993, 409–24.

Choudhuri, Arnab Rai. "Practicing Western Science Outside the West: Personal Observations on the Indian Scene." *Social Studies of Science* 15 (1985): 475–505.

Christen, Catherine. "At Home in the Field: Smithsonian Tropical Science Field Stations in the U.S. Panama Canal Zone and the Republic of Panama," *Americas* 58, no. 4 (April 2002): 537–75.

Committee on the Applications of Ecological Theory to Environmental Problems. *Ecological Knowledge and Problem-Solving: Concepts and Case Studies.* Washington, D.C.: National Academy Press, 1986.

Condit, Richard, Stephen Hubbell, and Robin Foster. "Mortality Rates of Two Hundred Five Neotropical Trees and Shrub Species and the Impact of a Severe Drought." *Ecological Monographs* 65, no. 4 (November 1995): 419–35.

Conference on the Future of the Smithsonian Institution. *Proceedings of the Conference on the Future of the Smithsonian Institution, February 11, 1927.* Baltimore: Lord Baltimore Press, 1927.

———. "Some Achievements during Eighty Years." In *Proceedings of the Conference on the Future of the Smithsonian Institution, February 11, 1927.* Baltimore: Lord Baltimore Press, 1927, 6–7.

Cooper, Mike. "U.S. Confirms West Nile Virus Caused N.Y. Deaths." Reuters Press, October 21, 1999.

Crichton, Michael. *Congo.* New York: Knopf, 1980.

Cronon, William. "The Trouble with Wilderness, or Getting Back to the Wrong Nature." In *Uncommon Ground,* ed. William Cronon, 69–90. New York: Norton, 1995.

D'Monte, Darryl. *Temples or Tombs: Industry versus Environment, Three Controversies.* New Delhi: Centre for Science and Environment, 1985.

Damodaran, V. K. "Silent Valley, Malabar, and the Power Paradox." *Economic and Political Weekly* 14, no. 40 (October 13, 1979): 1709–10.

Daniel, J. C. Review of *The Mountain Gorilla* by George Schaller. *Journal of the Bombay Natural History Society* 62, no. 1 (March 1965): 133–35.

———. "Unforgettable Sálim Ali." In *Sálim Ali's India,* ed. Ashok Kothari and B. F. Chhapgar, 14–18. Delhi: Oxford University Press, 1996.

———, ed. *A Century of Natural History.* Bombay: Bombay Natural History Society, 1983.

Daniel, J. C., and B. R. Grubh. "The Indian Wild Buffalo, *Bubalus bubalis,* in Peninsular India: A Preliminary Study." *Journal of the Bombay Natural History Society* 63, no. 1 (1966): 32–53.

Darlington, P. J. *Zoogeography: The Geographical Distribution of Animals.* New York: Wiley, 1957.

Das, Indraneil. "Animal Farm." *Seminar* 466 (June 1998): 50–52.

Dasgupta, Biplab. "The Environmental Debate: Some Issues and Trends." *Economic and Political Weekly* 13 (annual issue, February 1978): 385–400.

Dasmann, Raymond. *Planet in Peril: Man and the Biosphere Today.* New York: World Publishing, 1972.

———. "Toward a Biosphere Consciousness." In *The Ends of the Earth: Perspectives on Modern Environmental History,* ed. Donald Worster, 277–88. New York: Cambridge University Press, 1988.

Davis, Shelton. *Indigenous Views of Land and the Environment.* World Bank Discussion Papers, no. 188. Washington, D.C.: World Bank, 1993.

Davis, T. Anthony. "Selection of Nesting Trees and Frequency of Nest Visits by Baya Weaverbird." In *A Bundle of Feathers: Proffered to Sálim Ali for His Seventy-Fifth*

Birthday in 1971, ed. Dillon Ripley, 11–21. Delhi: Oxford University Press, 1978.

Denton, Peter. *Organizations That Help the World: World Wide Fund for Nature.* New Delhi: Orient Longman, 1996.

Deshmukh, B. G. Foreword to *Sálim Ali's India,* ed. Ashok Kothari and B. F. Chhapgar. Delhi: Oxford University Press, 1996.

Dhar, Uppeandra, ed. *Himalayan Biodiversity: Conservation Strategies.* Kosi, India: G. B. Pant Institute of Himalayan Environment and Development/Gyanodaya Prakashan Nainital, 1993.

Di Castri, Francesco. "The Programme on Man and the Biosphere (MAB)." In *Proceedings of the First International Congress of Ecology: Structure, Functioning, and Management of Ecosystems.* Wageningen, Netherlands: Centre for Agricultural Publishing and Documentation, 1974.

Di Castri, Francesco, F. W. G. Baker, and Malcolm Hadley, eds. *Ecology in Practice.* 2 vols. Part 1, *Ecosystem Management* (Excerpts from "Ecology in Practice: Establishing a Scientific Basis for Land Management," September 22–29, 1981, organized jointly by UNESCO and IUCN to mark the tenth anniversary of the MAB Programme. Part 2, *The Social Response.* Paris: UNESCO, 1984.

DiSilvestro, Roger. *Reclaiming the Last Wild Places: A New Agenda for Biodiversity.* New York: Wiley, 1993.

Diamond, Jared. "The Island Dilemma: Lessons of Modern Biogeographic Studies for the Design of Natural Reserves." *Biological Conservation* 7 (1975): 129–46.

Dugan, P. J., ed. *Wetland Conservation: A Review of Current Issues and Required Action.* Gland, Switzerland: IUCN/World Conservation Union, 1990.

Ehrlich, Paul. "Biodiversity and the Public Lands: Habitats in Crisis." *Wilderness,* Spring 1987, 12–15.

———. *The Population Bomb.* New York: Ballantine Books, 1968.

Ehrlich, Paul, David Dobkin, and Darryl Wheye. *The Birder's Handbook.* New York: Simon and Schuster, 1988.

Ehrlich, Paul, and Ann Ehrlich. *Extinction: The Causes and Consequences of the Disappearance of Species.* New York: Random House, 1981.

Eidsvik, Harold, et al. *The Biosphere Reserve and Its Relationship to Other Protected Areas.* Gland, Switzerland: IUCN, 1979.

Elliott, Hugh. *Second World Conference on National Parks: Yellowstone and Grand Teton National Parks, USA, September 18–27, 1972.* Morges, Switzerland: IUCN, 1974.

Emmel, Thomas. *Global Perspectives on Ecology.* Palo Alto, Calif.: Mayfield Publishing, 1977.

Enders, Robert. Review of *The Deer and the Tiger* by George Schaller. *American Scientist* 55, no. 3 (Fall 1967): 306A–308A .

Erwin, Terry. "Tropical Forest Canopies: The Last Biotic Frontier." *Bulletin of the Entomological Society of America* 29 (1983): 14–19.

Ewans, Martin. *Bharatpur: Bird Paradise.* London: H. F. and G. Witherby, 1989.

Featherstone, Drew, and John Cummings. "CIA Linked to 1971 Swine Virus in Cuba." *Washington Post,* January 9, 1977, A-2.

Finley, Hugh, et al., eds. *Lonely Planet Travel Survival Kit: India.* Oakland: Lonely Planet Publications, 1996.

Forster, Richard. *Planning for Man and Nature in National Parks: Reconciling Perpetuation and Use.* Morges, Switzerland: IUCN, 1973.

Fox, Michael, and A. J. T. Johnsingh. "Hunting and Feeding in Wild Dogs." *Journal of the Bombay Natural History Society* 72, no. 2 (1974): 321–26.

Frame, Bob, Joe Victor, and Yateendra Joshi, eds. *Biodiversity Conservation: Forests, Wetlands, and Deserts.* Proceedings of the Indo-British Workshops on Biodiversity, February, 1993, New Delhi. New Delhi: Tata Energy Research Institute, 1993.

Fuller, Steve. *Philosophy of Science and Its Discontents.* Boulder: Westview Press, 1989.

———. *Philosophy, Rhetoric, and the End of Knowledge: The Coming of Science and Technology Studies.* Madison: University of Wisconsin Press, 1993.

Futehally, Zafar. "A Portrait of Sálim Ali." In *A Bundle of Feathers: Proferred to Sálim Ali for His Seventy-Fifth Birthday in 1971,* ed. S. Dillon Ripley. Delhi: Oxford University Press, 1978.

Futehally, Zafar, et al. *Report on the Task Force for the Ecological Planning of the Western Ghats.* New Delhi: National Committee on Environmental Planning and Co-Ordination, Ministry of Science and Technology, Government of India, 1977.

Gadagkar, Raghavendra. "Biology in the Twenty-First Century: Back to Stamp Collection?" *Scampus,* Spring 1998, 5.

———. "Colony Founding in the Primitively Eusocial Wasp, *Ropalidia marginata* (Lep.) (Hymenoptera: Vespidae)." *Ecological Entomology* 20 (1995): 273–82.

———. "Do Social Wasps Choose Nesting Strategies Based on Their Brood Rearing Abilities?" *Naturwissenschaften* 84 (1997): 79–82.

———. "The Origin and Evolution of Caste Polymorphism in Social Insects." *Journal of Genetics* 76 (1997): 167–79.

———. "The Pains and Pleasures of Doing Ethology in India." In *Readings in Behaviour,* ed. R. Ramamurthi and Geethabali, 1–13. New Delhi: New Age International, 1996.

———. "Regulation of Worker Activity in a Primitively Eusocial Wasp, *Ropalidia marginata.*" *Behavioral Ecology* 6 (1995): 117–23.

———. *Survival Strategies.* Cambridge, Mass.: Harvard University Press, 1997.

———. "Western Scientists Set the Trends." *Down to Earth* 2, no. 7 (August 31, 1993): 45–47.

———. "Why the Definition of Eusociality is Not Helpful to Understand Its Evolution and What We Should Do about It." *Oikos* 70 (1994): 485–88.

Gadgil, Madhav. "Conservation: Where Are the People?" *The Hindu: Survey of the Environment 1998,* 101–37.

———. "Conservation of India's Living Resources through Biosphere Reserves." *Current Science* 51, no. 11 (June 3, 1982): 547–50.

———. "Dispersal: Population Consequences and Evolution." *Ecology* 52, no. 2 (Spring 1971): 253–61.

———. *Diversity: The Cornerstone of Life.* New Delhi: Vigyan Prasar, with Sanctuary Magazine, 1997.

———. *Establishment of Biosphere Reserves in India: Project Document 1; The Nilgiri Biosphere Reserve.* New Delhi: Indian National Man and the Biosphere Committee, Department of Environment, Government of India, 1980.

———. "In Love with Life." *Seminar 409: Our Scientists* (September 1993): 25–31.

———. "Sálim Ali (12 November 1896–20 June 1987)." *Current Science* 71, no. 9 (November 10, 1996): 686.

————. "Sálim Ali, Naturalist Extraordinaire: A Historical Perspective." *Journal of the Bombay Natural History Society* 75, supplement, 1980, i–v.

Gadgil, Madhav, and Ramachandra Guha. *Ecology and Equity: The Use and Abuse of Nature in Contemporary India.* New Delhi: Penguin, 1995.

————. *This Fissured Land: An Ecological History of India.* Delhi: Oxford University Press, 1992.

Gadgil, Madhav, and Kailash Malhotra. "Ecology Is for the People." *South Asian Anthropologist* 6, no. 1, 1985, 1–14.

Gadgil, Madhav, and V. M. Meher-Homji. "Role of Protected Areas in Conservation." In *Conservation for Productive Agriculture,* ed. V. L. Chopra and T. N. Khoshoo. New Delhi: Indian Council of Agricultural Research, 1986, 143–159.

Gadgil, Madhav, and R. Sukumar, eds. *Scientific Programme for the Nilgiri Biosphere Reserve: Report of a Workshop, Bangalore, March 7–8, 1986.* Bangalore: ENVIS Centre, Centre for Ecological Sciences, IISc, 1986.

Gadgil, Madhav, and V. D. Vartak. "Sacred Groves of India: A Plea for Continued Conservation." *Journal of the Bombay Natural History Society* 72, no. 2 (1974): 313–20.

Gaston, Anthony, Peter Garson, and Sanjeeva Pandey. "Birds Recorded in the Great Himalayan National Park, Himachal Pradesh, India." *Forktail* 9, 1993, 45–57.

Gee, E. P. "The Management of India's Wild Life Sanctuaries and National Parks." *Journal of the Bombay Natural History Society* 64, no. 2 (1967): 339–41.

————. Review of *The Deer and the Tiger* by George Schaller. *Journal of the Bombay Natural History Society* 64, no. 3 (1967): 530.

————. *The Wildlife of India.* London: Oxford University Press, 1964.

Geiger, Rudolf. *The Climate Near the Ground.* Trans. Scripta Technica, Inc. Cambridge, Mass.: Harvard University Press, 1965.

Glowka, Lyle, et al. *A Guide to the Convention on Biological Diversity: Environmental Policy and Law Paper no. 30.* Gland, Switzerland: IUCN, 1994.

Golding, Peter, and Phil Harris, eds. *Beyond Cultural Imperialism: Globalization, Communication, and the New International Order.* Thousand Oaks, Calif.: Sage Press, 1997.

Golley, Frank. *A History of the Ecosystem Concept: More than the Sum of the Parts.* New Haven: Yale University Press, 1993.

Gosse, Van. *Where the Boys Are: Cuba, Cold War America, and the Making of the New Left.* New York: Verso, 1993.

Gray, Andrew. *Between the Spice of Life and the Melting Pot: Biodiversity Conservation and Its Impact on Indigenous Peoples.* Copenhagen: International Work Group on Indigenous Affairs, document 70, 1991.

Greenberg, Brian, Subir Sinha, and Shubhra Gururani. "The 'New Traditionalist' Discourse of Indian Environmentalism." *Journal of Peasant Studies* 24, no. 3 (April 1997).

Gregg, William, Jr. "Some Thoughts on Biosphere Reserves: A Perspective From the U.S.A." In *Biosphere Reserves: Proceedings of the First National Symposium, Udhagamandalam, September 24–26, 1986.* New Delhi: Ministry of Environment and Forests, Government of India, 1987.

Gregg, William P., Jr., et al., eds. *Fourth World Wilderness Congress Worldwide Conservation: Proceedings of the Symposium on Biosphere Reserves, September 11–18, 1987, Estes Park, Colorado.* Atlanta: U.S. Department of the Interior, 1990.

Groombridge, Brian, ed. *Global Biodiversity: Status of the Earth's Living Resources: A Report Compiled by the World Conservation Monitoring Centre.* New York: Chapman and Hall, 1992.

Grove, Richard. *Green Imperialism: Colonial Expansion, Tropical Island Edens and the Origins of Environmentalism, 1600–1860.* New York: Cambridge University Press, 1995.

Grove, Richard, Vinita Damodaran, and Satpal Sangwan, eds. *Nature and the Orient: The Environmental History of South and Southeast Asia.* Delhi: Oxford University Press, 1998.

Grumbine, R. Edward. *Ghost Bears: Exploring the Biodiversity Crisis.* Washington, D.C.: Island Press, 1992.

———, ed. *Environmental Policy and Biodiversity.* Washington, D.C.: Island Press, 1994.

Guha, Ramachandra. "The Authoritarian Biologist and the Arrogance of Anti-Humanism: Wildlife Conservation in the Third World." *Ecologist* 27, no. 1 (January-February 1997): 14–20.

———. "Radical American Environmentalism and Wilderness Preservation: A Third World Critique." *Environmental Ethics* 11, no. 1 (Spring 1989): 71–83.

———. *Social Ecology.* Delhi: Oxford University Press, 1994.

———. *The Unquiet Woods: Ecological Change and Peasant Resistance in the Himalaya.* Delhi: Oxford University Press, 1989.

Gup, Ted. "The Smithsonian's Secret Contract: The Link between Birds and Biological Warfare." *Washington Post Magazine,* May 12, 1985, 10–17.

Gysel, L. W. "Borders and Openings of Beech-Maple Woodlands in Southern Michigan." *Journal of Forestry* 49 (1951): 13–19.

Haila, Yrjo. "Wilderness and Multiple Layers of Environmental Thought." *Environment and History* 3, 1997, 129–47.

Haraway, Donna. *Primate Visions.* New York: Routledge, 1989.

Harris, Robert, and Jeremy Paxman. *A Higher Form of Killing: The Secret Story of Chemical and Biological Warfare.* New York: Hill and Wang/Noonday Press, 1982.

Harroy, Jean-Paul. "A Century in the Growth of the 'National Park' Concept throughout the World." In *Second World Conference on National Parks: Yellowstone and Grand Teton National Parks, USA, September 18–27, 1972,* ed. Hugh Elliott. Morges, Switzerland: IUCN, 1974.

Hawkins, R. E., ed. *Encyclopedia of Indian Natural History: Centenary Publication of the Bombay Natural History Society 1883–1983.* Delhi: Oxford University Press, 1986.

Hellman, Geoffrey T. "Curator Getting Around." *New Yorker,* August 26, 1950, 31–49.

Henson, Pam. "Invading Arcadia: Women Scientists in the Field in Latin America, 1900–1950," *Americas* 58, no. 4 (April 2002): 577–600.

Hoch, Paul. Review: "Whose Scientific Internationalism?" *British Journal for the History of Science* 27 (September 1994): 345–49.

Home, R. W., and S. G. Kohlstedt, eds. *International Science and National Scientific Identity.* Boston: Kluwer Academic Publishers, 1991.

Hunter, Malcolm. *Wildlife, Forests, and Forestry: Principles of Managing Forests for Biological Diversity.* Englewood Cliffs, N.J.: Prentice-Hall, 1990.

India. Department of Environment. *National Wildlife Action Plan.* New Delhi: Department of Environment, Government of India, 1983.

————. Department of Science and Technology. MAB National Committee. *Research Projects 1975–78, MAB.* New Delhi: MAB National Committee, Department of Science and Technology, Government of India, 1978.

————. Department of Science and Technology. MAB National Committee. *Preliminary Inventory on Potential Areas for Biosphere Reserves.* New Delhi: MAB National Committee, Department of Science and Technology, Government of India, 1979.

————. Ministry of Environment and Forests. *Biosphere Reserves in India.* New Delhi: Ministry of Environment and Forests, 1989.

————. Ministry of Environment and Forests. *Biosphere Reserves: Proceedings of the First National Symposium, Udhagamandalam, September 24–26, 1986.* New Delhi: Ministry of Environment and Forests, 1987.

————. Ministry of Environment and Forests. *Wetlands, Mangroves, and Biosphere Reserves: Proceedings of the Indo-U.S. Workshop.* New Delhi: Ministry of Environment and Forests, 1989.

India. Planning Commission. *Seventh Five-Year Plan.* 1985.

————. *Sixth Five-Year Plan, 1980–85.* 1981.

————. *Sixth Five-Year Plan, 1980–85: A Framework.* 1980.

————. *Sixth Five-Year Plan, 1980–85: Mid Term Appraisal.* 1983.

Indian Board for Wild Life. *Task Force: Project Tiger: A Planning Proposal for Preservation of Tigers (Panthera tigris tigris Linn.) in India.* New Delhi: Ministry of Agriculture, 1972.

Indian Institute of Public Administration. *Biodiversity Conservation through Ecodevelopment: An Indicative Plan (Sponsored by the UNDP and the Ministry of Environment and Forests).* New Delhi: IIPA, 1994.

Instone, Lesley. *Science, Technology, and Western Domination: Some Aspects of Cultural Imperialism in the Third World.* Melbourne: Monash University, 1985.

International Biosphere Reserve Congress. *Conservation, Science, and Society: Contributions to the First International Biosphere Reserve Congress.* 2 vols. Minsk: UNESCO/UNEP, 1983.

IUCN (International Union for Conservation of Nature). *World Conservation Strategy.* Gland, Switzerland: IUCN, 1980.

————. Commission on National Parks and Protected Areas. *UN List of National Parks and Equivalent Reserves.* Gland, Switzerland: IUCN, 1980.

Jackson, Cecile. "Doing What Comes Naturally? Women and Environment in Development." *World Development* 21, no. 12, 1993, 1947–63.

James, Valentine Udoh. *Environmental and Economic Dilemmas of Developing Countries: Africa in the Twenty-First Century.* Westport, Conn.: Praeger, 1994.

Jayal, N. D. "Biosphere Reserves: Indian Approach." In International Biosphere Reserve Congress, *Conservation, Science, and Society: Contributions to the First International Biosphere Reserve Congress.* 2 vols. Minsk: UNESCO/UNEP, 1983.

Jha, Pramod Kumar, et al., eds. *Environment and Biodiversity in the Context of South Asia.* Kathmandu: Ecological Society, 1996.

Johnsingh, A. J. T. "Birdman Goes Darting." *Wildlife Institute of India Newsletter* (special commemorative issue, 1982–92), 1992, 17–18.

————. "Prey Selection in Three Large Sympatric Carnivores in Bandipur." *Mammalia* 56, no. 4 (December 1992): 517–26.

————. "Rajaji." *Sanctuary* 11 (1991): 14–25.

———. "The Threatened Gallery Forest of the River Tambiraparani, Mundantharai Wildlife Sanctuary, South India." *Biological Conservation* 47 (1989): 273–380.

———. "Vietnam Venture: The Primordial World of Sao La and Mang." *Frontline,* April 21, 1995, 94–97.

Johnsingh, A. J. T., and Justus Joshua. "Conserving Rajaji and Corbett National Parks: The Elephant as a Flagship Species." *Oryx* 28, no. 2 (April 1994): 135–40.

Jones, James. *Bad Blood.* New York: Free Press, 1981.

Joshi, N.V., and Madhav Gadgil. "On the Role of Refugia in Promoting Prudent Use of Biological Resources." *Theoretical Population Biology* 40, no. 2 (October 1991): 211–29.

Kahle, Jane Butler. "Women Biologists: A View and a Vision." *BioScience* 35, no. 4 (April 1985): 230–34.

Kamath, M.V. "Sponsoring Research in India: Pentagon Assailed." *Times of India,* August 16, 1969.

Kay, Charles E. "Yellowstone's Northern Elk Herd: A Critical Evaluation of the 'Natural Regulation' Paradigm." Ph.D. dissertation, Utah State University, 1990.

Kemf, Elizabeth. *WWF's Global Conservation Programme, 1997/98.* Gland, Switzerland: WWF-International, 1998.

Khoshoo, T. N. "Biodiversity in Developing Countries." In *Biodiversity, Science, and Development: Towards a New Partnership,* ed. F. Di Castri and T. Younes, 304–11. London: CAB International, 1996.

Kim, Ke Chung, and Robert Weaver, eds. *Biodiversity and Landscapes: A Paradox of Humanity.* New York: Cambridge University Press, 1994.

Kingsland, Sharon. *Modeling Nature: Episodes in the History of Population Biology.* 2d. ed. Chicago: University of Chicago Press, 1995.

Kissinger, Henry. *White House Years.* Boston: Little, Brown, 1979.

Kjekshus, Helge. *Ecology Control and Economic Development in East African History.* Berkeley: University of California Press, 1977.

Klemm, Cyrille de. *Biological Diversity Conservation and the Law: Legal Mechanisms for Conserving Species and Ecosystems.* Gland, Switzerland: IUCN, 1993.

Kothari, Ashish, P. Pande, S. Singh, and D. Variava. *Management of National Parks and Sanctuaries in India: A Status Report.* New Delhi: Environmental Studies Division Indian Institute of Public Administration, with the Ministry of Environment and Forests, 1989.

Kothari, Ashok, and B. F. Chhapgar, eds. *Sálim Ali's India.* Delhi: Oxford University Press, 1996.

Krattiger, Anatole, et al. *Widening Perspectives on Biodiversity.* Gland, Switzerland: IUCN, 1994.

Krausman, Paul R., and A. J. T. Johnsingh. "Conservation and Wildlife Education in India." *Wilderness Society Bulletin* 18, no. 3 (Fall 1990): 342–47.

Kuhn, Thomas. *The Structure of Scientific Revolutions.* Chicago: University of Chicago Press, 1962.

Kumar, Ajith, and Ravi Chellam. "Project Proposal: Impact of Fragmentation on the Biological Diversity of Rain Forest Small Mammals and Herpetofauna of the Western Ghats Mountains, South India." Manuscript, 1991.

Kumar, Deepak. *Science and the Raj, 1857–1905.* Oxford: Oxford University Press, 1995.

Kuppuram, G., and K. Kumudamani. *History of Science and Technology in India.* Vol. 12, *Environment and Ecology.* Delhi: Sundeep Prakashan, 1990.

Kwa, Chunglin. "Representations of Nature Mediating between Ecology and Science Policy: The Case of the International Biological Programme." *Social Studies of Science* 17 (1987): 413–42.

Latour, Bruno. *Science in Action.* Cambridge, Mass.: Harvard University Press, 1987.

———. *We Have Never Been Modern.* Cambridge, Mass.: Harvard University Press, 1993.

Lewis, Michael. Review of *A Carnival for Science* by Shiv Visvanathan. *Iowa Journal of Cultural Studies* 18 (1999): 119–21.

Lopez, Barry, and E. O. Wilson. "Ecology and the Human Imagination: Barry Lopez and E. O. Wilson." In *Writing Natural History: Dialogues with Authors,* ed. Edward Lueders, 7–36. Salt Lake City: University of Utah Press, 1989.

Lovejoy, Thomas, and David C. Oren. "The Minimum Critical Size of Ecosystems." In *Forest Island Dynamics in Man-Dominated Landscapes,* vol. 41 of *Ecological Studies,* ed. Robert L. Burgess and David M. Sharpe, 7–12. New York: Springer Verlag, 1981.

Lucas, P. H. C. *Protected Landscapes: A Guide for Policy-Makers and Planners.* New York: Chapman and Hall for IUCN, 1992.

MAB Program. *Biosphere Reserves in Action: Case Studies of the American Experience.* Washington, D.C.: U.S. MAB Program, Department of State Publications, 1995.

MacArthur, Robert, and E. O. Wilson. "An Equilibrium Theory of Insular Zoogeography." *Evolution* 17 (December 1963): 373–87.

———. *The Theory of Island Biogeography.* Princeton: Princeton University Press, 1967.

MacKenzie, John. *The Empire of Nature: Hunting, Conservation, and British Imperialism.* New York: Manchester University Press, 1988.

MacKinnon, John, and Kathy MacKinnon, eds. *Review of the Protected Areas System in the Indo-Malayan Realm.* Gland, Switzerland: IUCN, 1986.

MacKinnon, John, et al., eds. *Managing Protected Areas in the Tropics (based on the Workshops on Managing Protected Areas in the Tropics World Congress on National Parks, Bali, Indonesia, October, 1982).* Gland, Switzerland: IUCN, 1986.

Martin, Emily. *Flexible Bodies: The Role of Immunity in American Culture from the Days of Polio to the Age of AIDS.* Boston: Beacon Press, 1994.

Maslow, Jonathan. *Footsteps in the Jungle: Adventures in the Scientific Exploration of the American Tropics.* Chicago: Ivan R. Dee, 1996.

Mayr, Ernst. "The Zoogeographic Position of the Hawaiian Islands." *Condor* 45, 1943, 45–48.

McClintock, Anne. *Imperial Leather: Race, Gender, and Sexuality in the Colonial Conquest.* New York: Routledge, 1995.

McClure, Elliott. *Bird Banding.* Pacific Grove, Calif.: Boxwood Press, 1984.

———. *Stories I Like to Tell: An Autobiography.* Camarillo, Calif.: Elliott McClure, 1995.

McNeely, Jeffrey, and David Pitt, eds. *Culture and Conservation: The Human Dimension in Environmental Planning.* Dover, N.H.: Croom Helm for IUCN, 1985.

McNeely, Jeffrey, et al., eds. *People and Protected Areas in the Hindu Kush–Himalaya: Proceedings of the International Workshop on the Management of National Parks and Protected Areas in the Hindu Kush–Himalaya,* May 6–11, 1985, Kathmandu. Kathmandu: King Mahendra Trust for Nature Conservation and the International Centre for Integrated Mountain Development, 1985.

Meena, B. L., and Shruti Sharma. *Keoladeo National Park, Bharatpur, Rajasthan.* Jaipur: Rajasthan Forest Department, 1998.

Mehmet, Ozay. *Westernizing the Third World: The Euro-Centricity of Economic Development Theories.* New York: Routledge, 1995.

Menon, M. G. K., et al. *Ecological Aspects of the Silent Valley: Report of the Joint Committee Set Up by the Government of India and the Government of Kerala.* New Delhi: Department of Environment, Government of India, 1982.

Menon, Vivek, Raman Sukumar, and Ashok Kumar. *A God in Distress: Threats of Poaching and the Ivory Trade to the Asian Elephant in India.* Bangalore: Asian Elephant Conservation Centre, 1997.

Murali, K. S., and Raman Sukumar. "Reproductive Phenology of a Tropical Dry Forest in Mudumalai, Southern India." *Journal of Ecology* 82, 1994, 759–67.

Myers, Norman. *The Sinking Ark: A New Look at the Problem of Disappearing Species.* New York: Pergamon Press, 1979.

Nair, Sathis Chandran. *The Southern Western Ghats: A Biodiversity Conservation Plan.* New Delhi: India National Trust for Art and Cultural Heritage, 1991.

Nanda, Meera. "The Epistemic Charity of the Social Constructivist Critics of Science and Why the Third World Should Refuse the Offer." In *A House Built on Sand: Exposing Postmodernist Myths about Science,* ed. Noretta Koertge, 286–311. New York: Oxford University Press, 1998.

Nandy, Ashish, ed. *Science, Hegemony, and Violence: A Requiem for Modernity.* Delhi: Oxford University Press, 1988.

Nash, Roderick. *Wilderness and the American Mind.* 3d ed. New Haven: Yale University Press, 1982.

National Research Council. *Science and the Endangered Species Act.* Ed. Michael Clegg et al. Washington, D.C.: National Academy Press, 1995.

Needham, Joseph, LuGwei-Djen, et al. *The Hall of Heavenly Records: Korean Astronomical Instruments and Clocks, 1360–1780.* Cambridge: Cambridge University Press, 1986.

Odum, Eugene. *Fundamentals in Ecology.* Philadelphia: Saunders, 1953.

Oehser, Paul. *The Smithsonian Institution.* 2d rev. ed. Boulder: Westview Press, 1983.

Orlove, Benjamin. "Ecological Anthropology." *Annual Review of Anthropology* 9, 1980, 235–73.

Painter, Michael, and William Durham, eds. *The Social Causes of Environmental Destruction in Latin America.* Ann Arbor: University of Michigan Press, 1995.

Palladino, Paulo. *Entomology, Ecology, and Agriculture: The Making of Scientific Careers in North America, 1885–1985.* Amsterdam: Harwood Academic Press, 1996.

Palladino, Paulo, and Michael Worboys. "Science and Imperialism." *Isis* 84 (1993): 91–102.

Panwar, H. S., and B. K. Mishra. "Rajaji National Park: Real Issues, Problems, and Prospects." *Himachal Times,* October 6, 1994, 7.

Parameswaran, M. P. "Significance of Silent Valley." *Economic and Political Weekly* 14, no. 27 (July 7, 1979): 1117–19.

Park, Edwards. "Secretary S. Dillon Ripley Retires after Twenty Years of Innovation." *Smithsonian* 15, no. 6 (September 1984): 76–87.

Pearson, Graham S. "BTWC Security Implications of Human, Animal, and Plant Epidemiology." *Strengthening the Biological Weapons Convention.* Briefing paper no. 23,

Report of the NATO Advanced Research Workshop, Cantacuzino Institute, Bucharest, June 3–5, 1999. Brussels: NATO, 1999.

Perrings, Charles, et al., eds. *Biodiversity Loss: Economic and Ecological Issues.* New York: Cambridge University Press, 1995.

Petras, J. "Cultural Imperialism in the Late Twentieth Century." *Journal of Contemporary Asia* 23, no. 2 (1993): 139–48.

Pianka, Eric. *Evolutionary Ecology.* 5th ed. New York: HarperCollins College Publishers, 1994.

Pimm, S. L., G. J. Russell, J. L. Gittleman, and T. M. Brooks. "The Future of Biodiversity." *Science* 269 (1995): 347–50.

Porter, Gareth, and Janet Brown. *Global Environmental Politics.* San Francisco: Westview Press, 1991.

Porter, Theodore. *Trust in Numbers: The Pursuit of Objectivity in Science and Public Life.* Princeton: Princeton University Press, 1995.

Poti, P. S. *Rajaji: The Indian People's Tribunal on Environment and Human Rights: Preliminary Report on the Rajaji National Park.* Bombay: Indian People's Tribunal, 1994.

Prater, S. H. *The Book of Indian Animals.* 2d ed. Delhi: Oxford University Press, 1971 [1948].

Pratt, Mary Louise. *Imperial Eyes: Travel Writing and Transculturation.* New York: Routledge, 1992.

Preston, F. W. "The Canonical Distribution of Commonness and Rarity." Parts 1 and 2. *Ecology* 43 (1962): 185–215, 410–32.

Price, Jennifer. *Flight Maps: Adventures with Nature in Modern America.* New York: Basic Books, 1999.

Pyenson, Lewis. "Cultural Imperialism and Exact Sciences Revisited." *Isis* 84 (1993): 107.

Quammen, David. *The Song of the Dodo: Island Biogeography in an Age of Extinction.* New York: Simon and Schuster, 1996.

Raj, Kapil. "Images of Knowledge, Social Organization, and Attitudes to Research in an Indian Physics Department." *Science in Context* 2, no. 2 (1988): 317–49.

Rajiv, B. "U.S. and India: New Directions and Their Context." *Economic and Political Weekly* 12, no. 46 (November 5, 1977): 1913.

Rangarajan, Mahesh. *Fencing the Forest: Conservation and Ecological Change in India's Central Provinces, 1860–1914.* Delhi: Oxford University Press, 1996.

———. "Five Nature Writers: Corbett, Anderson, Ali, Sankhala, and Krishnan." In *Environment and Wildlife: Five Essays, Research in Progress Papers, History and Society,* 3d series, no. 29, 14–31. New Delhi: Nehru Memorial Museum and Library, March 1998.

———. *India's Wildlife History.* Delhi: Permanent Black, 2001.

———. "The Politics of Ecology: The Debate on Wildlife and People in India, 1970–95." *Economic and Political Weekly* (September 1996): 2391–2409.

———. *Troubled Legacy: A Brief History of Wildlife Preservation in India.* Occasional Paper, New Delhi: Nehru Memorial Museum and Library, 1988.

Ranjitsinh, M. K. "Autonomy Document for Wildlife Institute of India." Internal memo no. 23–17/84-FRY (WL). New Delhi: Department of Environment, Forests, and Wildlife (Wild Life Division), 1986.

Reed, Nathaniel P. "How Well has the United States Managed Its National Park System? The Application of Ecological Principles to Park Management." In Hugh Elliott, ed., *Second World Conference on National Parks: Yellowstone and Grand Teton National Parks, USA, September 18–27, 1972*. Morges, Switzerland: IUCN, 1974.

Regis, Ed. *The Biology of Doom: The History of America's Secret Germ Warfare Project*. New York: Holt, 1999.

Reid, W. V. "How Many Species Will There Be?" In *Tropical Deforestation and Species Extinction*, ed. T. C. Whitmore and J. A. Sayer. London: Chapman and Hall, 1992.

Ripley, S. Dillon. "A New Race of Nightjar from Ceylon." *Bulletin of the British Ornithologists' Club* 65 (June 20, 1945): 40–41.

———. "The Smithsonian's Role in U.S. Cultural and Environmental Development." *BioScience* 36, no. 3 (March 1986): 155.

———. *A Synopsis of the Birds of India and Pakistan, Together with Those of Nepal, Sikkim, Bhutan, and Ceylon*. Bombay: Bombay Natural History Society, 1961.

———. "The View from the Castle." *Smithsonian* 14 (December 1983): 10.

———, ed. *A Bundle of Feathers: Proffered to Sálim Ali for His Seventy-Fifth Birthday in 1971*. Delhi: Oxford University Press, 1978.

Ross, Andrew. *The Chicago Gangster Theory of Life: Nature's Debt to Society*. New York: Verso, 1994.

———, ed. *Science Wars*. Durham: Duke University Press, 1996.

Rossiter, Margaret. *Women Scientists*. Baltimore: Johns Hopkins University Press, 1982.

Roy, P. C. *Indo-U.S. Economic Relations*. New Delhi: Deep and Deep Publications, 1986.

Saberwal, Vasant. *Pastoral Politics: Shepherds, Bureaucrats, and Conservation in the Western Himalaya*. Delhi: Oxford University Press, 1999.

Saberwal, Vasant, James Gibbs, Ravi Chellam, and A. J. T. Johnsingh. "Lion–Human Conflict in the Gir Forest, India." *Conservation Biology* 8, no. 2 (June 1994): 501–7.

Saberwal, Vasant, Mahesh Rangarajan, and Ashish Kothari. *People, Parks, and Wildlife: Towards Coexistence*. New Delhi: Orient Longman, 2001.

Saharia, V. B. "The Building Blocks of Wildlife Education in India." *WII Newsletter*, October 1992 (commemorative issue, 1982–91): 7–9.

Sainath, P. *Everybody Loves a Good Drought: Stories from India's Poorest Districts*. New Delhi: Penguin, 1996.

Saldanha, Cecil J. *A Select Bibliography on the Environment of Karnataka*. Bangalore: Centre for Taxonomic Studies, St. Joseph's College, 1993.

Sale, Kirkpatrick. *The Green Revolution: The American Environmental Movement, 1962–1992*. New York: Hill and Wang, 1993.

Sangwan, Satpal. *Science, Technology, and Colonisation: An Indian Experience, 1757–1857*. Delhi: Anamika Prakashan, 1991.

Sankhala, Kailash. "Livestock Grazing in India's National Parks." In *People and Protected Areas in the Hindu Kush–Himalaya: Proceedings of the International Workshop on the Management of National Parks and Protected Areas in the Hindu–Kush Himalaya, 6–11 May, 1985*, ed. J. A. McNeely, J. W. Thorsell, S. R. Chalise. Kathmandu: King Mahendra Trust for Nature Conservation and International Centre for Integrated Mountain Development, 1985, 55–58.

———. *Tiger! The Story of the Indian Tiger*. New York: Simon and Schuster, 1977.

Saunders, Denis, Richard Hobbs, and Chris Margules. "Biological Consequences of Ecosystem Fragmentation: A Review." *Conservation Biology* 5, no. 1 (March 1991): 19–20.

Schaden, Herman. "Bird Project a Coverup for War Test, NBC Says." *Washington (D.C.) Evening Star*, February 4, 1969, A-1.

Schaller, George. *The Deer and the Tiger: A Study of Wildlife in India*. Chicago: University of Chicago Press, 1967.

———. *Golden Shadows, Flying Hooves*. New York: Random House, 1973.

———. *The Mountain Gorilla: Ecology and Behavior*. Chicago: University of Chicago Press, 1963.

———. *The Year of the Gorilla*. Chicago: University of Chicago Press, 1964.

Schulze, Ernst-Detlef, and Harold Mooney, eds. *Biodiversity and Ecosystem Function*. Berlin: Springer-Verlag, 1993.

Shaffer, Mark. "Minimum Population Sizes for Species Conservation." *BioScience* 31, no. 2 (February 1981): 131–34.

Shahabuddin, Ghazala, Ranjit Lab, and Pratibha Pande. *Small and Beautiful: Sultanpur National Park*. New Delhi: Kalpavriksh, 1995.

Shankar, Alaka. *Indira Priyadarshini*. New Delhi: Children's Book Trust, 1986.

Shiva, Vandana. *Monocultures of the Mind: Perspectives on Biodiversity and Biotechnology*. Dehradun: Natraj Publishers, 1993.

———, ed. *Biodiversity: Social and Ecological Perspectives*. London: Zed Books, 1991.

Shiva, Vandana, and Vanaja Ramprasad. *Cultivating Diversity: Biodiversity Conservation and the Politics of the Seed*. Dehradun: Research Foundation for Science, Technology, and Natural Resource Policy, 1993.

Siddiqi, Javed. *World Health and World Politics*. Columbia: University of South Carolina Press, 1995.

Simberloff, Daniel. "The Contribution of Population and Community Biology to Conservation Science." *Annual Review of Ecology and Systematics* 19 (1988): 473–511.

Simberloff, Daniel, and L. G. Abele. "Island Biogeography and Conservation Strategy and Limitations." *Science* 193 (1976): 1032.

Singh, Arjan. *Tiger Haven*. Delhi: Oxford University Press, 1998.

Singh, Baldev, ed. *Jawaharlal Nehru on Science and Society: A Collection of His Writings and Speeches*. New Delhi: Nehru Memorial Museum and Library, 1988.

Singh, Pramod. *Indian Environment*. Proceedings of national conference organized by Institute of Environmental Sciences, "Chitrakut Gramoday Vishwavidyalay Chitrakut." New Delhi: Ashish Publishing House, 1992.

Singh, Samar. *India's Action Plan for Wildlife Conservation and Role of Voluntary Bodies*. BNHS Centenary Seminar, 6–10 December, 1983. New Delhi: WWF-India, 1984.

Singh, Shekar. *Biodiversity Conservation through Ecodevelopment: Planning and Implementation Lessons from India*. South-South Cooperation Programme on Environmentally Sound Socio-Economic Development in the Humid Tropics, Working Papers, no. 21. Paris: UNESCO, 1997.

Smith, Henry Nash. *Virgin Land*. Cambridge, Mass.: Harvard University Press, 1950.

Soule, Michael. "What Is Conservation Biology?" *BioScience* 35, no. 11 (December 1985): 727–34.

Soule, Michael, and Daniel Simberloff. "What Do Genetics and Ecology Tell Us about the Design of Nature Reserves?" *Biological Conservation* 35 (1986): 19–40.

Soule, Michael, and B. A. Wilcox, eds. *Conservation Biology: An Evolutionary-Ecological Perspective.* Sunderland, Mass.: Sinauer Press, 1980.

Spence, Mark. *Dispossessing the Wilderness: Indian Removal and the Making of the National Parks.* New York: Oxford University Press, 1999.

Spillett, J. Juan. "The Ecology of the Lesser Bandicoot Rat in Culcutta." Calcutta: Bombay Natural History Society/Johns Hopkins University Center for Medical Research and Training, 1968.

————. "General Wild Life Conservation Problems in India." *Journal of the Bombay Natural History Society* 63, no. 3 (1966): 616–29.

Sreberny-Mohammadi, Annabelle. "The Many Cultural Faces of Imperialism." In *Beyond Cultural Imperialism: Globalization, Communication, and the New International Order,* ed. Peter Golding and Phil Harris, 49–68. Thousand Oaks, Calif.: Sage Press, 1997.

Srinivasan, Kannan. "The Environmentalists: Another View." *Economic and Political Weekly* 15, no. 11 (April 12, 1980): 693–96.

Stoessel, W. J., et al. Appendix E in *Report of the Chemical Warfare Commission.* Washington, D.C.: U.S. Government Printing Office, 1985.

Subrahmanyam, K. V. "Environment of Silent Valley." *Economic and Political Weekly* 15, no. 40 (October 4, 1980): 1651–52.

Sukumar, Raman. *The Asian Elephant: Ecology and Management.* New York: Cambridge University Press, 1989.

————. "Dynamics of a Tropical Deciduous Forest: Population Changes (1988 through 1993) in a Fifty-Hectare Plot at Mudumalai, Southern India." In *Forest Biodiversity Research, Monitoring, and Modeling: Conceptual Background and Old World Case Studies,* ed. F. Dallmeier and James A. Comiskey. Delhi: Parthenon Publishing, 1997.

————. *Elephant Days and Nights: Ten Years with the Indian Elephant.* New Delhi: Oxford University Press, 1994.

————. "Elephant-Man Conflict in Karnataka." In *Karnataka–State of Environment Report, 1984–85,* ed. C. J. Saldanha, 46–59. Bangalore: Centre for Taxonomic Studies, 1986.

————. "The Management of Large Mammals in Relation to Male Strategies and Conflicts with People." *Biological Conservation* 55 (1991): 93–102.

————. "Minimum Viable Populations for Asian Elephant Conservation." In *A Week with Elephants,* ed. J. C. Daniel and J. Datye, 279–88. New Delhi: Oxford University Press, 1995.

————. "The Nagerhole Tiger Controversy." *Current Science* 59, no. 23 (December 10, 1990): 1213–16.

Sunquist, Fiona. *Tiger Moon.* Chicago: University of Chicago Press, 1988.

Swanson, Timothy. *The Economics and Ecology of Biodiversity Decline: The Forces Driving Global Change.* New York: Cambridge University Press, 1995.

————. *The International Regulation of Extinction.* New York: New York University Press, 1994.

Takacs, David. *The Idea of Biodiversity: Philosophies of Paradise.* Baltimore: Johns Hopkins University Press, 1996.

Thapar, Valmik. "Fatal Links." *Seminar* 466 (June 1998): 59–69.

Thorsell, J. W., ed. *Conserving Asia's Natural Heritage: The Planning and Management of Protected Areas in the Indomalayan Realm.* Proceedings of the Twenty-Fifth Working

Session of IUCN's Commission on National Parks and Protected Areas, Corbett National Park, India, February 4–8, 1985. Gland, Switzerland: IUCN, 1985.

Tomlinson, John. *Cultural Imperialism.* Baltimore: Johns Hopkins University Press, 1991.

UNEP (United Nations Environment Programme). *Convention on Biological Diversity.* Paris: Environmental Law and Institutions Programme Activities Centre, 1992.

UNESCO (United Nations Educational, Scientific, and Cultural Organization). *Biosphere Reserves: The Seville Strategy and the Statutory Framework of the World Network.* Paris: UNESCO, 1996.

———. *Intergovernmental Conference of Experts on the Scientific Basis for Rational Use and Conservation of the Resources of the Biosphere, Final Report: SC/MD/9.* Paris: UNESCO, January 6, 1969.

———. *International Co-Ordinating Council of the Programme on Man and the Biosphere (MAB): First Session, Final Report.* Paris: UNESCO, November 9–19, 1971.

———. *International Co-Ordinating Council of the Programme on Man and the Biosphere (MAB): Thirteenth Session, Final Report.* Paris: UNESCO, 1995.

United States. Department of State. *Humid Tropical Forests: AID Policy and Program Guidance.* Washington, D.C.: Department of State Memorandum, 1985.

Vijayan, Lalitha. *Keoladeo National Park, Rajasthan.* New Delhi: WWF-India, 1994.

Vijayan, V. S. *Keoladeo National Park Ecology Study.* Bombay: Bombay Natural History Society, 1987.

———. *Keoladeo National Park Ecology Study, 1980–1990 Summary.* Bombay: Bombay Natural History Society, 1991.

Visvanathan, Shiv. *A Carnival for Science.* New Delhi: Oxford University Press, 1996.

———. "A Celebration of Difference: Science and Democracy in India." *Science* 280 (1998): 42.

Waters, Malcolm. *Globalization.* New York: Routledge, 1995.

Webb, H. E., et al. "Leukemia and Neoplastic Process Treated with Langat and Kyasanur Forest Disease Viruses: A Clinical and Laboratory Study of Twenty-Eight Patients." *British Medical Journal,* January 29, 1966, 258–66.

Wemmer, Chris, and K. Ullas Karanth. "Reflections on Fieldwork Abroad: Attitudes and Latitudes: Observations and Platitudes." *Grapevine* 39 (March 2001): 3–4.

Whitmore, T. C., and J. A. Sayer, eds. *Tropical Deforestation and Species Extinction.* London: Chapman and Hall, 1992.

WII (Wildlife Institute of India). *Annual Report, 1986–87.* Dehradun: Wildlife Institute of India, 1987.

———. *Annual Report, 1987–88.* Dehradun: Wildlife Institute of India, 1988.

———. *Annual Report, 1992–93.* Dehradun: Wildlife Institute of India, 1993.

———. *Annual Report, 1994–95.* Dehradun: Wildlife Institute of India, 1995.

———. *Wildlife Institute of India: A Profile.* Dehradun: Wildlife Institute of India, 1993.

Wilson, E. O. "The Biological Diversity Crisis." *BioScience* 35, no. 11 (December 1985): 700–706.

———. *Biophilia.* Cambridge, Mass.: Harvard University Press, 1984.

———. *Consilience.* New York: Vintage Books, 1998.

———. *The Insect Societies.* Cambridge, Mass.: Belknap Press of Harvard University Press, 1971.

———. *Naturalist.* Washington, D.C.: Island Press, 1994.

————. *Sociobiology: The New Synthesis.* Cambridge, Mass.: Belknap Press of Harvard University Press, 1975.

————, ed. *Biodiversity: National Forum on BioDiversity.* Washington, D.C.: National Academy Press, 1986.

Wolpert, Stanley. *A New History of India.* 4th ed. New York: Oxford University Press, 1993.

Work, Telford H., and H. Trapido. "Summary of Preliminary Report on Investigation of the Virus Research Centre on an Epidemic Disease Affecting Forest Villagers and Wild Monkeys of Shimoga District." *Indian Journal of Medical Science* 11 (1957): 340–41.

World Research Institution. "Land Conversion." <http://www.wri.org/wri/wr-98–99/landconv.htm>.

World Resources Institute. *Global Biodiversity Strategy: Guidelines for Action to Save, Study and Use Earth's Biotic Wealth Sustainably and Equitably.* New York: WRI/IUCN/UNEP, 1992.

Worthington, E. B., ed. *The Evolution of IBP.* Cambridge: Cambridge University Press, 1975.

WWF-India (World Wide Fund for Nature). *Participatory Management Planning for Keoladeo National Park: Report on the Participatory Rural Appraisal (PRA) Workshop held at the Keoladeo National Park, Bharatpur, November 20-December 5, 1995.* New Delhi: WWF-India, 1997.

————. *Regional Cooperation in South Asia: The WWF Framework.* New Delhi: WWF-India, 1997.

————. *Role of NGOs in Wildlife Conservation: A Discussion Paper.* New Delhi: WWF-India, 1987.

WWF-UK. *Conservation of Tropical Forests and Biodiversity: The Role of the United Kingdom.* Gland, Switzerland: WWF, 1996.

Index